HONORS CALCULUS

HONORS CALCULUS

Charles R. MacCluer

PRINCETON UNIVERSITY PRESS
Princeton and Oxford

Copyright © 2006 by Princeton University Press
Published by Princeton University Press, 41 William Street,
Princeton, New Jersey 08540
In the United Kingdom: Princeton University Press,
3 Market Place, Woodstock, Oxfordshire OX20 1SY

Library of Congress Cataloging-in-Publication Data
MacCluer, C. R.
Honors calculus / Charles R. MacCluer.
   p. cm.
Includes bibliographical references and index.
ISBN-13: 978-0-691-12533-6 (cloth : alk. paper)
ISBN-10: 0-691-12533-3 (cloth : alk. paper)
1. Calculus—Textbooks. I. Title.
QA303.2.M33 2006
515—dc22   2005054519

British Library Cataloging-in-Publication Data is available

This book has been composed in ITC Stone Serif and
ITC Stone Sans

Printed on acid-free paper. ∞

pup.princeton.edu

Printed in the United States of America

10 9 8 7 6 5 3 2 1

This book is dedicated to my son Joshua.

Contents

Preface   xi
Acknowledgments   xiii

1 Functions on Sets
1.1 Sets   1
1.2 Functions   2
1.3 Cardinality   5
Exercises   6

2 The Real Numbers
2.1 The Axioms   12
2.2 Implications   14
2.3 Latter-Day Axioms   16
Exercises   16

3 Metric Properties
3.1 The Real Line   19
3.2 Distance   20
3.3 Topology   21
3.4 Connectedness   22
3.5 Compactness   23
Exercises   27

4 Continuity
4.1 The Definition   30
4.2 Consequences   31
4.3 Combinations of Continuous Functions   33
4.4 Bisection   36
4.5 Subspace Topology   37
Exercises   38

5 Limits and Derivatives
5.1 Limits   41
5.2 The Derivative   43
5.3 Mean Value Theorem   46
5.4 Derivatives of Inverse Functions   48
5.5 Derivatives of Trigonometric Functions   50
Exercises   53

6 **Applications of the Derivative**

6.1 Tangents	60
6.2 Newton's Method	63
6.3 Linear Approximation and Sensitivity	65
6.4 Optimization	66
6.5 Rate of Change	67
6.6 Related Rates	68
6.7 Ordinary Differential Equations	69
6.8 Kepler's Laws	71
6.9 Universal Gravitation	73
6.10 Concavity	76
6.11 Differentials	79
Exercises	80

7 **The Riemann Integral**

7.1 Darboux Sums	89
7.2 The Fundamental Theorem of Calculus	91
7.3 Continuous Integrands	92
7.4 Properties of Integrals	94
7.5 Variable Limits of Integration	95
7.6 Integrability	96
Exercises	97

8 **Applications of the Integral**

8.1 Work	100
8.2 Area	102
8.3 Average Value	104
8.4 Volumes	105
8.5 Moments	106
8.6 Arclength	109
8.7 Accumulating Processes	110
8.8 Logarithms	110
8.9 Methods of Integration	112
8.10 Improper Integrals	113
8.11 Statistics	115
8.12 Quantum Mechanics	117
8.13 Numerical Integration	118
Exercises	121

9 **Infinite Series**

9.1 Zeno's Paradoxes	134
9.2 Convergence of Sequences	134
9.3 Convergence of Series	136

9.4 Convergence Tests for Positive Series   138
9.5 Convergence Tests for Signed Series   140
9.6 Manipulating Series   142
9.7 Power Series   145
9.8 Convergence Tests for Power Series   147
9.9 Manipulation of Power Series   149
9.10 Taylor Series   151
Exercises   154

References   163
Index   165

This book is for the honors calculus course that many universities offer to well-prepared entering students. As a rule, all of these students will have had an earlier calculus course where they have been exposed to many of the standard ideas and have been drilled in the standard calculations. These students are therefore primed for an exceptional experience, yet most universities deliver the same material again in much the same style. Instead, this book can be used to catapult these exceptional students to a much higher level of understanding.

To the Student

Mathematics is more than calculations. It is also about constructing abstract structures that can make laborious calculations unnecessary. And it is about *understanding* with a precision and clarity unequaled by other disciplines. In this book you will be exposed to how mathematics is practiced by mathematicians. It is my hope that you may leapfrog the usual twice relearning of calculus as a junior and as a graduate student by learning it correctly the first time (although I suspect your presence in honors calculus signals that you have studied some calculus before).

I will assume your algebra skills are excellent. I will also assume that you are sensitive to the precision of language used in science; that is your key to success in our journey—ponder until you understand *exactly* what is being said.

Look carefully at definitions and results. What function does each word play within the statement? Can you think of an example that satisfies all but one of the criteria but where the definition or result fails? Can you think of a simple example where it holds? Much of the understanding and recalling of mathematics lies in amassing a large personal zoo of examples that you can trot out to test definitions and results.

Like any discipline, much drill is necessary for mastery, but here, both computational and conceptual drill is required. Here are suggestions for study: Take scratch notes during lectures, then reconstruct and expand these notes later into your own personal text. Next, read the corresponding section of the text, and as you read, follow along with a pen and pad checking all details. Finally, attack the exercises. Spend a portion of every study time reviewing. There seems to be no shortcut around this time-consuming study technique.

Like you would in French class, repeatedly practice the italicized epigrams of the results out loud, like "continuous functions vanish on

closed sets" or "the continuous image of a compact set is compact."
Repeat these mantras often.

To the Instructor

Given the correct clientele it is possible to teach calculus right the first
time. Tiresome repetition and incremental advance is not suitable for
the honors student. It is most important not to bore such students.
They must be challenged.

All of your students will have been exposed to the routine calculus
material. If you begin trudging through the same material, albeit at a
higher level, they will tune out and spend their intellectual energies
elsewhere. The approach and material of this book will be new to your
students and very demanding because of its abstraction. But they will
master it.

Details are often thrown into the exercises. You may present this
detail or assign it. The overarching objective is to reveal the flow of the
ideas, and so not all proofs need be presented, but it is crucial that the
students believe that all proofs are accessible given the requisite time.
Although the analysis is built up axiomatically, examples are drawn
from the assumed past experience of the students, and only later, as
the analysis builds, is this background material verified.

The usual repetitious drill in routine calculations is missing from
this book. You may need to augment it with problems from a standard
calculus book. Honors students pick up techniques from working only
one or two standard problems; the usual dreary repetition is not necessary. The goal has been to construct a framework upon which you
may tailor your course to your tastes and to the talent present.

By design, the writing style is brief, breezy, and *au courant*. The intent
is to charm and engage the young clientele.

Solutions are available to instructors from the publisher upon
request. Additions and corrections to the text will be updated at http://
www.math.msu.edu/~maccluer/HonorsCalculus.

Acknowledgments

A number of students helped shape this book. They suffered through the early write-ups and typos and often suggested valuable improvements. I especially thank J. Ryan Brunton, Umar Farooq, Emma Beth Hummel, and Daniel C. Laboy.

I thank my colleagues Wm. C. Brown, Michael W. Frazier, Peter Magyar, John D. McCarthy, Jacob Plotkin, Clark J. Radcliffe, Christel Rotthaus, William T. Sledd, Clifford E. Weil, and Vera Zeidan. At one time or another, they all provided invaluable help.

The reviewers of this book offered many helpful suggestions. For their insights I thank Adrian Banner, Princeton University, and one unnamed reviewer.

Most of all I am indebted to Paul Halmos, my erstwhile teacher and mentor, whose mathematical style permeates this book.

Charles R. MacCluer
maccluer@msu.edu

HONORS CALCULUS

1

Functions on Sets

We take as a working definition that *mathematics is the study of functions on sets*. In this chapter we take up the primitive notions of sets, functions from one set to another, and injective, surjective, and bijective functions. Sets are classified as finite and infinite, countable and uncountable. All areas of mathematics use these fundamental concepts. This is the core of mathematics.

1.1 Sets

Suppose X is a collection (*set*) of objects (*points*) generically denoted by x. A *subset* S of X is a collection consisting of some of the objects of X, in symbols, $S \subset X$. For any two subsets S and T of X, their *intersection* $S \cap T$ is the set of all objects x in both S and T. Of course it is quite possible that S and T share no common points, in which case we say their intersection is the *empty set* $\emptyset$, and we write $S \cap T = \emptyset$.

The *union* $S \cup T$ is the set of all objects x in either S or T.[1] Two subsets S and T are *equal*, in symbols, $S = T$, if[2] they consist of exactly the same objects.

Subsets A, B, C, ... of X under "cap" and "cup" enjoy an arithmetic of sorts:

Theorem A. (Boolean algebra)

$$
\begin{aligned}
A \cap B &= B \cap A & A \cup B &= B \cup A \\
A \cap (B \cap C) &= (A \cap B) \cap C & A \cup (B \cup C) &= (A \cup B) \cup C \\
A \cap A &= A & A \cup A &= A \\
A \cap X &= A & A \cup \emptyset &= A \\
A \cap \emptyset &= \emptyset & A \cup X &= X
\end{aligned}
$$
$$
\begin{aligned}
A \cup (B \cap C) &= (A \cup B) \cap (A \cup C) \\
A \cap (B \cup C) &= (A \cap B) \cup (A \cap C).
\end{aligned}
\tag{1.1}
$$

---

[1] The word "or" is used in mathematics in the inclusive sense—either one or the other or both.

[2] We continue the age-old practice when defining terms of using "if" when we really should use "exactly when" or "if and only if." See [Euclid].

Proof. Such set-theoretic formulas are proved by tedious "point-chasing" arguments. For instance, let us prove the first of the distributive laws,

$$A \cup (B \cap C) = (A \cup B) \cap (A \cup C).$$

Let x be a point of the left-hand side, in symbols, $x \in A \cup (B \cap C)$. Then (Case 1) $x \in A$ or (Case 2) $x \in B \cap C$, that is, $x \in B$ and $x \in C$. If (Case 1) $x \in A$, then certainly $x \in A \cup B$ and $x \in A \cup C$, and hence $x \in (A \cup B) \cap (A \cup C)$. On the other hand, if (Case 2) $x \in B$ and $x \in C$, then certainly $x \in (A \cup B)$ and $(A \cup C)$, again giving $x \in (A \cup B) \cap (A \cup C)$.

Conversely, suppose x is an arbitrary point of the right side, that is, $x \in (A \cup B) \cap (A \cup C)$. Then $x \in A \cup B$ and $x \in A \cup C$. But this means $x \in A$ or B and $x \in A$ or C. If (Case 1) $x \in A$, then certainly $x \in A \cup (B \cap C)$. On the other hand, if (Case 2) x is not in A, in symbols, $x \notin A$, then nevertheless $x \in B$ and C, again giving that $x \in A \cup (B \cap C)$. Obtain practice with this boring but important proof method by completing exercise 1.1.

Some of the tedium can be avoided by observing the *duality* of cup and cap—the identities of theorem A remain unchanged when one interchanges cap with cup and $\emptyset$ with X. Exercise 1.3 exploits this duality insight when coupled with the following theorem B.

Theorem B. (the De Morgan laws)

$$(A \cap B)' = A' \cup B' \quad \text{and} \quad (A \cup B)' = A' \cap B', \qquad (1.2)$$

where the prime denotes the operation of *complementation*: the set S' is the set of all points of X not in S.

Proof. Exercise 1.2.

1.2 Functions

Definition. Let X and Y be two sets. A *function (mapping)* from X to Y, in symbols

$$f : X \longrightarrow Y, \qquad (1.3)$$

is a rule that assigns to each point x in X exactly one point y in Y, that is, $f(x) = y$. The set X is called the *domain* of the function f.

Warning. The function f is the rule, not its *graph,* the set G of all ordered pairs $(x, f(x))$ with $x \in X$.

Example 1. Suppose X is the set of people on Earth and Y the set of real numbers **R** (to be defined carefully in Chapter 2). Let f be the rule that assigns to each person their height in meters.

Example 2. Let $X = Y = \mathbf{N}$ be the set of all *natural numbers* $\mathbf{N} = \{1, 2, 3, \ldots\}$. Consider the rule f that assigns to each $x \in X$ the number of distinct divisors of x.

Example 3. A function may be given as a table:

x	a	b	c	d	e
$f(x)$	3	4	1	1	2

In this example, $X = \{a, b, c, d, e\}$ and Y is some set containing the symbols 1, 2, 3, 4. The rule assigns a to 3, b to 4, c to 1, and so on. Note that the function repeats in value; the value 1 is taken on twice—the function is not injective (one-to-one).

Definition. A function $f : X \longrightarrow Y$ is *injective* if no value y is taken on more than once, that is, for all $x_1, x_2 \in X$,

$$f(x_1) = f(x_2) \quad \text{implies} \quad x_1 = x_2. \tag{1.4}$$

For example, the rule of example 1 assigning people to their height is injective, since with unlimited precision it is certain that no two people are exactly the same height.

Example 4. The function $f : \mathbf{R} \longrightarrow \mathbf{R}$ given by the rule $f(x) = x^2$ is two-to-one, since $f(x) = f(-x)$, and is hence not injective. Horizontal lines cut its graph more than once.

Definition. The *range* $f(X)$ of a function $f : X \longrightarrow Y$ is the subset of Y of all values taken on by f as x runs through all of X. That is, $y \in f(X)$ if and only if $y = f(x)$ for some $x \in X$. See figure 1.1.

In Example 4, the range of $f(x) = x^2$ is the set of all nonnegative real numbers. The range in example 3 is $f(X) = \{1, 2, 3, 4\}$.

Definition. A function $f : X \longrightarrow Y$ is *surjective* (*onto*) if the range of f is all of Y, that is, $f(X) = Y$; every object in Y occurs at least once among the values of f on X.

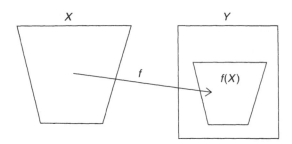

Figure 1.1 The range of the function f is the subset of Y consisting of the values of f.

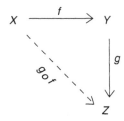

Figure 1.2 The composition $g \circ f$ is the function obtained by operating by g on the result of operating by f.

Definition. A function $f : X \longrightarrow Y$ that is both injective and surjective is called *bijective* (*one-to-one, onto*). Looking backward through such a bijective function yields (exercise 1.9) the bijective *inverse* function

$$f^{-1} : Y \longrightarrow X. \tag{1.5}$$

One last notion.

Definition. Suppose $f : X \longrightarrow Y$ and $g : Y \longrightarrow Z$. Then the function

$$h : X \longrightarrow Z$$

given by the rule $h(x) = g(f(x))$ is called the *composition* of f followed by g. We write $h = g \circ f$. See figure 1.2. Note that composition is associative (Exercise 1.13).

Theorem C. The composition of injective functions is injective. The composition of surjective functions is surjective. The composition of bijective functions is bijective.

Proof. Exercise 1.14.

With these simple notions and classifications in our toolkit, we are ready to attack some significant issues.

1.3 Cardinality

When do two sets X and Y contain exactly the "same number" of points? The answer is clear if both sets are finite—you merely count the number of points in each set and compare the count. But what if the sets are infinite? For that matter, what does it mean that X is an infinite set? What is the concept of infinity?

Definition. The sets X and Y are said to be of the *same cardinality* if there exists a bijective function from X onto Y. We indicate that X and Y have the same cardinality with the notation

$$|X| = |Y|. \tag{1.6}$$

Example 5. The rule

$$f(x) = \frac{e^x}{1 + e^x} \tag{1.7}$$

maps the entire real line $\mathbf{R}$ bijectively onto the interval $(0, 1)$, and so

$$|\mathbf{R}| = |(0, 1)|. \tag{1.8}$$

(Exercise 1.10.) So even though the interval $(0, 1)$ is finite in length and a (very) proper subset of $\mathbf{R}$, it nevertheless contains exactly the "same number" of elements as its superset $\mathbf{R}$. This is the defining property of infinity.

Definition. The set X is said to be *infinite* if there exists an injective function $f : X \longrightarrow X$ onto a proper subset of itself, where $f(X) \neq X$. The set X is *finite* if it is not infinite, that is, no injective, nonsurjective function exists.

Corollary. (the pigeonhole principle) Every injective function f from a finite set X to itself is surjective.

Remark. We write $|X| \leq |Y|$ when there exists an injective map from X into Y. It is surprisingly difficult to prove that (exercise 1.40)

$$|X| \leq |Y| \quad \text{and} \quad |Y| \leq |X| \quad \text{implies} \quad |X| = |Y|. \tag{1.9}$$

This result is known as the Schröder-Bernstein theorem.

Definition. The (ring of) *integers* is the set

$$\mathbf{Z} = \{\ldots, -3, -2, -1, 0, 1, 2, 3, \ldots\}, \tag{1.10}$$

the smallest subset of $\mathbf{R}$ closed under addition that contains both 1 and -1. (More about this definition in chapter 2.)

Theorem D. The integers have smaller cardinality than the reals, in symbols, $|\mathbf{Z}| < |\mathbf{R}|$. That is, there is an (obvious) injection from $\mathbf{Z}$ into $\mathbf{R}$, but no injection from $\mathbf{R}$ into $\mathbf{Z}$. Thus there are at least two distinct infinite cardinalities.

Proof. Exercise 1.31.

Any set with the same cardinality as $\mathbf{Z}$ is said to be a *countably infinite* set. All other infinite sets are said to be *uncountably infinite*.

Question. Do there exist sets X with

$$|\mathbf{Z}| < |X| < |\mathbf{R}|? \tag{1.11}$$

Exercises

1.1   Practice point chasing by proving the first five identities for cap in theorem A.

1.2   Prove the De Morgan laws (1.2).

1.3   Deduce the first five identities for cap (union) in theorem A by taking the complement of the first five identities for cup (intersection) and applying the De Morgan laws (theorem B).

For example, since we have $A' \cup B' = B' \cup A'$, taking complements gives $(A' \cap B')' = A'' \cup B'' = A \cup B = (B' \cup A')' = B'' \cap A'' = B \cap A$.

1.4   The set $A \setminus B$ is the set of all points in A that do not lie in B. Prove (by chasing points) or disprove (by example) that $A \setminus (B \cup C) = (A \setminus B) \cup (A \setminus C)$.

1.5   Is $(A \setminus B) \cup (B \setminus A) = A \cup B$?

1.6   Prove or disprove (by example):
          a. $A \cap B = A \cap C$ implies $B = C$.
          b. $A \cup B = A \cup C$ implies $B = C$.

1.7   What is the graphical interpretation of the statement
       "$f : \mathbf{R} \longrightarrow \mathbf{R}$ is injective"?

1.8   What is the graphical interpretation of the statement
       "$f : \mathbf{R} \longrightarrow \mathbf{R}$ is surjective"?

1.9   Prove carefully that given a bijective function $f : X \longrightarrow Y$, there exists
       exactly one bijective function $g : Y \longrightarrow X$ prescribed by $g(f(x)) = x$.
       Thus *a bijective function possesses a bijective inverse function.*

1.10   Argue that the function of (1.7) is a bijective map of $\mathbf{R}$ onto $(0, 1)$.

1.11   Give an example of a function $f : \mathbf{R} \longrightarrow \mathbf{R}$ that is not injective, that
       in fact takes on each of its values infinitely often.
           Suggestion: Examine $f(x) = \sin x$. What is its range?

1.12   Give an example of a function $f : \mathbf{R} \longrightarrow \mathbf{R}$ that is surjective, and in
       fact takes on every real value infinitely often.

1.13   Prove that composition is associative, that is, if $f : X \longrightarrow Y$,
       $g : Y \longrightarrow Z$, and $h : Z \longrightarrow W$, then $h \circ (g \circ f) = (h \circ g) \circ f$.

1.14   Prove theorem C.

1.15   Calculate the number N of possible functions $f : X \longrightarrow Y$ from a set
       X of cardinality m into a set Y of cardinality n.
           Answer: $N = n^m$.

1.16   Calculate the number N of bijective functions f from a set X of
       cardinality n onto itself.
           Answer: $N = n!$.

1.17   Calculate the number N of injective functions from a set X of
       cardinality m into a set Y of cardinality n.
           Answer: $N = 0$ when $m > n$. Otherwise, $N = n!/(n - m)!$.

1.18   Count the number of surjective functions from a set of five elements
       onto a set of two elements.

1.19* Calculate the number N of surjective functions from a set X of
       cardinality m onto a set Y of cardinality n.
           Answer: $N = 0$ when $m < n$. Otherwise,

$$N = n^m - \binom{n}{n-1}(n-1)^m + \binom{n}{n-2}(n-2)^m - \cdots .$$

1.20   Composition is rarely commutative. For example, $\sin x^2 \neq (\sin x)^2$.
        Give an example of two functions f, g from a finite set X to itself
        where $f \circ g \neq g \circ f$.

1.21   Prove $|\mathbf{Z}| = |\mathbf{N}|$.

1.22*   Deduce from the pigeonhole principle that a surjective function f
        from a finite set X onto itself is injective.
        Outline: Consider the composition of the set-valued injection g
        given by $g(x) = f^{-1}(\{x\})$ followed by an injection h that selects one
        element from each preimage $g(x)$.

1.23   Seven people enter an elevator and each in turn presses one of the
        five floor buttons. What is the probability that all buttons will have
        been pushed?

1.24   **(Lord Russell's paradox)** Allowing sets of unlimited size leads to
        disaster. Let Ω be the set of all sets that do not contain themselves
        as elements. Answer this question: Does Ω contain itself as an
        element?

1.25*   Prove that the set of *rational numbers* $\mathbf{Q}$, the set of all fractions
        $q = a/b$, where $a, b \in \mathbf{Z}$, $b \neq 0$, is countably infinite. That is, show
        $|\mathbf{Z}| = |\mathbf{Q}|$.
        Cantor's Diagonalization: List all positive fractions (with repeats) in
        an array:

$$
\begin{array}{cccccc}
1/1 & 2/1 & 3/1 & 4/1 & 5/1 & 5/2 \\
1/2 & 2/2 & 3/2 & 4/2 & \bullet & \bullet \\
1/3 & 2/3 & 3/3 & \bullet & \bullet & \bullet \\
1/4 & 2/4 & \bullet & \bullet & \bullet & \bullet \\
1/5 & \bullet & \bullet & \bullet & \bullet & \bullet
\end{array}
$$

        Count the entries starting in the upper left hand corner using the
        above serpentine pattern:

1.26   Prove that if $g \circ f$ is injective, then so is f.

1.27   Prove that if $g \circ f$ is surjective, then so is g.

1.28   Suppose $f : X \longrightarrow Y$. For any subset S of Y, the symbol $f^{-1}(S)$ denotes
        the *preimage of S under* f, that is, the subset of X consisting of all x

mapped by f into S. Show that for any two subsets S, T of Y,

$$f^{-1}(S \cap T) = f^{-1}(S) \cap f^{-1}(T)$$
$$f^{-1}(S \cup T) = f^{-1}(S) \cup f^{-1}(T).$$

1.29   In contrast to exercise 1.28, are the following statements always true?

$$f(S \cap T) = f(S) \cap f(T),$$
$$f(S \cup T) = f(S) \cup f(T).$$

1.30   Show that for any set X, no matter how large in cardinality, there exists a set Y of even larger cardinality, that is, there exists an injection from X into Y but no injection from Y into X. Thus there is a never-ending hierarchy of infinities.
          Outline: Let $Y = \mathcal{P}(X)$ be the *power set* of X, the set of all subsets of X. There is clearly the injection from X into Y given by mapping each x to the singleton $\{x\}$. Assume however that there exists a bijection g from X onto Y. Let X_0 be the subset of X consisting of all points $x \in X$ not belonging to their image under g. Find the preimage x_0 of X_0 under g. Neither $x_0 \in X_0$ nor $x \notin X_0$ can obtain.

1.31   Show that $|\mathbf{Z}| < |\mathbf{R}|$.
          Outline: It is enough to show that the reals in $(0, 1)$ are not countable. With some care, such reals have unique decimal expansions, not all of which can be enumerated, since one may construct a decimal expansion that differs in the first place from the first, differs in the second place from the second, and so on.

1.32   Let Y be the power set of X. In your opinion, does there exist a function $g : Y \longrightarrow X$ with the property that $g(y) \in y$ for all $y \in Y$? That is, can you select an element from each subset of a family of subsets?

1.33   Given two sets X and Y, then $X \times Y$ denotes the *cartesian product* of X with Y, the set of all ordered pairs $X \times Y = \{(x, y); \, x \in X, \, y \in Y\}$. For example, $\mathbf{R}^2 = \mathbf{R} \times \mathbf{R}$ is the familiar Euclidean plane. Sketch the subset of $\mathbf{R}^2$ that is the cartesian product $(0, 1) \times \mathbf{R}$.

1.34   Show that the operations of addition and multiplication in $\mathbf{R}$ are functions of the type $f : \mathbf{R} \times \mathbf{R} \longrightarrow \mathbf{R}$.

1.35   Let T denote the unit circle given by $x^2 + y^2 = 1$ in $\mathbf{R}^2$ and let I be the set of all reals $0 < x < 1$ in $\mathbf{R}$. Sketch the cartesian product $T \times I$ in $\mathbf{R}^3$.

1.36   Prove that the collection G of all bijective functions from a set X onto itself forms a *group*, that is,

> *Axiom I:* For all $f, g, h \in G$, $f \circ (g \circ h) = (f \circ g) \circ h$.
> *Axiom II:* There is an element $e \in G$ with the property $e \circ f = f \circ e = f$ for all $f \in G$.
> *Axiom III:* For each $f \in G$ there exists an element $g \in G$ for which $f \circ g = g \circ f = e$.

This group G is called the *symmetric group on the symbols of X*. See [Artin].

1.37   More concretely than in exercise 1.36, let us consider the symmetric group S_4 on four letters, that is, the twenty-four *permutations* of the digits 1, 2, 3, 4. Each of these permutations can be written in a clever shorthand as a product of disjoint *cycles*. For example, the bijective map (permutation) f that maps according to the rule

x	1	2	3	4
$f(x)$	2	1	4	3

can be simply written as (12)(34), which compactly indicates that $1 \mapsto 2 \mapsto 1$ and $3 \mapsto 4 \mapsto 3$. Fixed digits are not mentioned. For example, the permutation

x	1	2	3	4
$f(x)$	2	4	3	1

is given by (124). The identity permutation $f(i) = i$ is written simply as (1).
Write down all twenty-four elements of the group S_4 in cycle notation. Find pairs of permutations that do not commute. Find permutations that are self-inverse.

1.38   Construct the 6×6 multiplication (composition) table for the symmetric group S_3 on the three digits 1, 2, 3.

1.39   Prove that every subset of a finite set is finite.

1.40*   Prove the **Schröder-Bernstein theorem:** $|X| \leq |Y|$ and $|Y| \leq |X|$
        implies $|X| = |Y|$.
        Outline: [Halmos] Suppose both $f : X \to Y$ and $g : Y \to X$ are
        injective. A point w in either X or Y is said to be a *descendant* of a
        point z in either X or Y if w is the eventual image of z under
        alternating applications of f and g; we also say then that z is an
        ancestor of w. Some points may be orphans. Partition X into three
        disjoint subsets: the set X_X of descendants of orphans in X, the set
        X_Y of descendants of orphans in Y, and the set X_∞ of points with
        infinitely many ancestors. Likewise, partition $Y = Y_X \cup Y_Y \cup Y_\infty$.
        Then $f : X_X \to Y_X$, $g : Y_Y \to X_Y$, and $f : X_\infty \to Y_\infty$ are all bijective.
        Piece together f, g^{-1}, and f on X_X, X_Y, and X_∞ respectively to
        obtain a bijection from X onto Y.

1.41*   Prove that $|\mathbf{R}| = |\mathcal{P}(\mathbf{N})|$, where $\mathcal{P}(\mathbf{N})$ is the power set of $\mathbf{N}$.
        Outline: Note that a subset S of $\mathbf{N}$ can be uniquely specified by a
        sequence of zeros and ones by placing a 1 in the nth place of the
        sequence when $n \in S$, 0 if not. Such a sequence of zeros and ones
        represents a real number in the interval $[0, 1]$ written base 2. Thus
        there exists a surjection from the power set of $\mathbf{N}$ onto $[0, 1]$.

2

The Real Numbers

In this chapter we carefully build up from axioms the most important of all analytic objects, the field of real numbers $\mathbf{R}$.

2.1 The Axioms

Axiom I. The set of real numbers $\mathbf{R}$ forms a *field* under addition and multiplication. That is, for $x, y, z \in \mathbf{R}$ and the two distinct elements $0, 1 \in \mathbf{R}$, we have

$$x + (y + z) = (x + y) + z \qquad\qquad x \cdot (y \cdot z) = (x \cdot y) \cdot z$$
$$x + y = y + x \qquad\qquad x \cdot y = y \cdot x$$
$$0 + x = x \qquad\qquad 1 \cdot x = x$$

For each x there is a y such that $x + y = 0$ For each $x \neq 0$ there is a y such that $x \cdot y = 1$

$$x \cdot (y + z) = x \cdot y + x \cdot z$$

From this short list of field properties follows all the familiar laws of arithmetic; see exercises 2.1–2.8 and in particular the definitions of the symbols $-x$ and x^{-1}.

Axiom II. The real numbers $\mathbf{R}$ form an *ordered* field. That is, there is a subset P of $\mathbf{R}$ of *positive numbers* with the properties

$x, y \in P$ implies $x + y \in P$,
$x, y \in P$ implies $x \cdot y \in P$, and
for every $x \in \mathbf{R}$, exactly one of $x \in P$, $x = 0$, or $-x \in P$ holds.

Corollary. The square of any nonzero real number is positive.

Proof. Choose any $x \neq 0$. If x is positive, then so is $x \cdot x = x^2$. If x is negative, that is, $-x$ is positive, then $(-x) \cdot (-x) = (-x)^2 = x^2$ is positive—see exercises 2.3 and 2.4.

Definition. The *inequality* $a < b$ (or equivalently $b > a$) means $b - a$ is positive. The inequality $a \leq b$ (or equivalently $b \geq a$) means $b - a$ is either positive or 0.

Theorem A. For all $a, b, c \in \mathbf{R}$,

$$a < b \quad \text{and} \quad b < c \quad \text{implies} \quad a < c. \tag{2.1a}$$

$$\text{For any } x, \quad a < b \quad \text{implies} \quad a + x < b + x. \tag{2.1b}$$

$$a < b \quad \text{and} \quad p > 0 \quad \text{implies} \quad p \cdot a < p \cdot b. \tag{2.1c}$$

Proof. Exercise 2.10.

All the standard inequalities now follow. See the exercises.

Definition. The set $\mathbf{N}$ of *natural numbers* is the intersection of all subsets of $\mathbf{R}$ that are closed under addition and contain the number 1. Clearly $\mathbf{N} \supset \{1, \; 1 + 1, \; 1 + 1 + 1, \; \ldots\}$. [In fact, equality holds given axiom III (exercise 2.13).]

Axiom III. The natural numbers $\mathbf{N}$ are *well ordered*: Every nonempty subset S of $\mathbf{N}$ contains a *least element,* to wit, a point s_0 of S such that $s_0 \leq s$ for all $s \in S$. Axiom III is proof by induction in disguise—see exercise 2.16.

Axiom IV. The real numbers are *complete*: Every nonempty subset S of $\mathbf{R}$ that is bounded from above possesses a *supremum.*

A set S is said to be *bounded from above* by b if $x \leq b$ for all $x \in S$. The *supremum* of such a bounded set is an upper bound b that does not exceed any other upper bound—it is the *least upper bound*. This least upper bound is denoted by the symbol $\sup S$ and is clearly unique.

Example 1. Consider the (open) interval $S = (0, 1)$, the set of all reals x with $0 < x < 1$. This set is clearly bounded above by say 2. Therefore S possesses a supremum; in this case obviously $\sup S = 1$.

Example 2. Far more telling is the situation for a set of partial decimal expansions of, for instance, the famous real number π:

$$S = \{3, \; 3.1, \; 3.14, \; 3.141, \; 3.1415, \; 3.14159, \; 3.141592, \; \ldots\}. \tag{2.2}$$

The set S is bounded above by say 3.2, also by 3.1416, also by 3.141593, and so on. The value π is defined as the supremum of these partial decimal expansions.

This is the power of the calculus—the ability to make precise and to predict the result of an infinite number of calculations.

2.2 Implications

Theorem B. (Archimedean property) For any real x and positive real p, there exists a unique integer n so that

$$np \leq x < (n+1)p. \tag{2.3}$$

Proof. As in most proofs, we reduce to a special case at the crux of the matter: Note that by dividing (2.3) through by p, it is sufficient to prove the result for the case $p = 1$, that is,

$$n \leq x < n+1. \tag{2.4}$$

Suppose x exceeds all integers. Then by completeness (axiom IV), the set $\mathbf{Z}$ would possess a supremum, say z_0, a least upper bound to all integers. But then there must be an integer n with $z_0 - 1 < n \leq z_0$, for otherwise $z_0 - 1$ would be an even smaller upper bound than z_0 for $\mathbf{Z}$. But adding 1 to the inequality $z_0 - 1 < n$ yields the contradiction $z_0 < n + 1$.

Thus there are integers $n > x$. Let S be the set of all such integers. By well ordering (exercise 2.17) there must be a smallest such integer n_0 giving that

$$n_0 - 1 \leq x < n_0. \tag{2.5}$$

Take $n = n_0 - 1$ in (2.4). It is commonplace to denote the greatest integer $n \leq x$ by the symbol $[x]$.

Corollary A. For any positive real number ϵ, no mater how small, there is a natural number n whose reciprocal is even smaller, that is,

$$0 < \frac{1}{n} < \epsilon. \tag{2.6}$$

There are arbitrarily small positive rational numbers.

Proof. Find a natural number n with $n - 1 \leq 1/\epsilon < n$. Reciprocate.

Corollary B. Between every two distinct real numbers is a rational.

Proof. Suppose $x < y$. Choose $n \in \mathbf{N}$ with $1/n < y - x$. By (2.3) with $p = 1/n$ find $m \in \mathbf{Z}$ such that

$$\frac{m}{n} \leq x < \frac{m+1}{n}.$$

But the right-hand side term

$$\frac{m+1}{n} = \frac{m}{n} + \frac{1}{n} \leq x + \frac{1}{n} < x + y - x = y,$$

that is,

$$x < \frac{m+1}{n} < y. \tag{2.7}$$

Thus *every real can be approximated to any degree of accuracy by rational numbers.*

These observations open floodgates. For example,

Fact. Every positive real number a possesses exactly one positive N-th root $a^{1/N}$ for each natural number N.

Proof. We may as well assume $a > 1$, since $(1/a)^{1/N} = 1/a^{1/N}$. Let S be the set of all positive reals x whose N-th power does not exceed a. Note $1 \in S$. Moreover, S is clearly bounded above by (say) a itself, since for any $1 < x \in S$ with $x^N \leq a$, we have $1 < x < x^2 < x^3 < \cdots < x^N \leq a$ (exercise 2.18). Thus S possesses the least upper bound $s = \sup S$. But then for all natural numbers k, because s is the least upper bound of S,

$$\left(s - \frac{s}{k}\right)^N < a < \left(s + \frac{s}{k}\right)^N, \tag{2.8}$$

hence

$$\left(1 - \frac{1}{k}\right)^N < \frac{a}{s^N} < \left(1 + \frac{1}{k}\right)^N, \tag{2.9}$$

and so for some $p, q > 0$ (exercise 2.19),

$$1 - \frac{p}{k} < \frac{a}{s^N} < 1 + \frac{q}{k}, \tag{2.10}$$

that is,

$$-\frac{p}{k} < \frac{a}{s^N} - 1 < \frac{q}{k}. \tag{2.11}$$

But $a/s^N - 1$ is a constant independent of k that by (2.11) is caught between numbers arbitrarily close to 0, hence $a/s^N = 1$. This root s is the only possible positive N-th root of a by exercise 2.21.

2.3 Latter-Day Axioms

Logicians of the twentieth century established that there are two highly desirable properties that are *independent* of the four axioms displayed in §2.1 and the standard set theory touched upon in chapter 1. This means that one can adjoin one or both of these properties or their negations to familiar mathematics to obtain four divergent but consistent schools of mathematics. In spite of isolated firestorms of dissent, mathematicians eventually reached a consensus: Mathematics is far more interesting and rich with the addition of the following two axioms.

Axiom V. (continuum hypothesis) There exists no cardinality between that of the natural numbers and the reals. In symbols, for any set X,

$$|\mathbf{N}| \le |X| \le |\mathbf{R}| \quad \text{implies} \quad |X| = |\mathbf{N}| \quad \text{or} \quad |X| = |\mathbf{R}|. \qquad (2.12)$$

Axiom VI. (axiom of choice) Given any collection Ω of nonempty subsets of the set X, there exists a function $f : \Omega \longrightarrow X$ such that for each $S \in \Omega$, we have $f(S) \in S$. Informally, one can choose one point from each set S in Ω.

Although it defies intuition, Zermelo's axiom of choice is neither obvious nor "true;" it cannot be deduced from the classical axioms. Interestingly, the *generalized continuum hypothesis*, which denies existence to cardinalities between any set and its power set, implies the axiom of choice. The issue igniting the most heat was Zermelo's proof that any set, in particular $\mathbf{R}$, can be well ordered—one can define an ordering $\prec$ such that any nonempty subset contains a smallest element. This ferment and its resolution is one of the highpoints of twentieth-century mathematics. For further reading on roads taken and abandoned, consult [Hofstadter], [Cantor], [Halmos], and [Weyl]. Do a web search on the names Cantor, Zermelo, Frankel, Frege, and Russell.

Exercises

2.1   Prove the additive *cancellation law*: $x + y = x + z$ implies $y = z$.
      Hint: Add an additive inverse of x to both sides.

2.2   Deduce from exercise 2.1 that additive inverses are unique, that is, for each $x \in \mathbf{R}$ that there is one and only one element $y \in \mathbf{R}$ so that $x + y = 0$. This unique element y is denoted by $-x$.

2.3   Prove that $0 \cdot x = 0$ for all $x \in \mathbf{R}$. Deduce that $-x = (-1) \cdot x$.

2.4   Prove that $(-1) \cdot (-1) = 1$.

2.5   Prove the multiplicative *cancellation law*: $x \neq 0$ and $x \cdot y = x \cdot z$
        implies $y = z$.

2.6   Deduce from exercise 2.5 that multiplicative inverses are unique, that
        is, for each $x \neq 0$ in $\mathbf{R}$ there is one and only one element $y \in \mathbf{R}$ so that
        $x \cdot y = 1$. This unique element y is denoted sometimes by x^{-1}, other
        times by $1/x$.

2.7   Prove that when $x \neq 0$ in $\mathbf{R}$, we have $1/(1/x) = x$.

2.8   Prove that when $a, b, c, d \in \mathbf{R}$ and $b \cdot d \neq 0$, then
$$\frac{a}{b} + \frac{c}{d} = \frac{a \cdot d + b \cdot d}{b \cdot d}.$$
        Hint: By definition, $a/b = a \cdot (1/b)$.

2.9   Which field properties hold for $\mathbf{Z}$? For $\mathbf{Q}$? For $\mathbf{N}$?

2.10   Prove theorem A.

2.11   Prove that $a \leq b$ and $b \leq a$ implies $a = b$.

2.12   Show that the (field) of complex numbers $\mathbf{C}$ cannot be ordered.
            Hint: Square the imaginary number i.

2.13   Prove that $\mathbf{N} = \{1, \ 1+1, \ 1+1+1, \ \ldots\}$.
            Outline: First show $\mathbf{N}$ consists of positive numbers. Then let n_0 be
            the least of all the elements of $\mathbf{N}$ not of the form $n = 1 + 1 + \cdots + 1$
            (n times). Strike it out.

2.14   Using the symbols 0, 1, 2, 3, 4, write down an addition and
        multiplication table between these symbols to obtain a field $\mathbf{Z}_5$ of five
        elements. Can this field be ordered?

2.15   Prove that every finite set X of real numbers can be subscripted with
        consecutive natural numbers, that is, $X = \{x_1, x_2, \ldots, x_n\}$, so that
        $x_1 < x_2 < \cdots < x_n$.

2.16   Prove that for all natural numbers n, we have
$$1 + 2 + 3 + \cdots + n = \frac{n(n+1)}{2}.$$

Outline: Proceed by *induction*. Suppose there is a nonempty set S of natural numbers n where the result fails. By well ordering, S has a least element m. Since the result clearly holds for $n = 1$, we must have $m > 1$. Moreover, by the minimality of m, the result holds for $n = m - 1$, that is,

$$1 + 2 + 3 + \cdots + m - 1 = \frac{(m-1)m}{2}.$$

Add m to both sides to obtain that the result holds for $n = m$, a contradiction. Thus S is empty, and the result holds for all natural numbers.

2.17   The (ring of) *integers* **Z** is the intersection of all subsets of **R** that are closed under addition and subtraction, and contain 1. Prove that $\mathbf{Z} = \{\ldots, -1 - 1 - 1, -1 - 1, -1, 0, 1, 1 + 1, 1 + 1 + 1, \ldots\}$.

2.18   Prove that for any positive real $x > 1$ we have $1 < x < x^2 < x^3 < \cdots$.

2.19   Deduce the existence of $p, q > 0$ in (2.10).

2.20   Prove that if $0 < x < y$, then $1/x > 1/y$.

2.21   Prove that if $0 < x < y$, then $0 < x^N < y^N$ for any natural number N. Thus $f(x) = x^N$ is an increasing function on the positive reals.

2.22   Prove that when $a, b, c > 0$, then

$$a < b \quad \text{implies} \quad \frac{a}{b} < \frac{a+c}{b+c}.$$

2.23   Prove that $\sqrt{2} \notin \mathbf{Q}$.

2.24   Prove that every nonempty set $S \subset \mathbf{R}$ that is bounded from below possesses a greatest lower bound (called the *infimum* of S denoted by the symbol inf S).
        Hint: Consider $-S = \{x; \; x = -s, \; s \in S\}$.

2.25   **(Project)** Prepare a historical account of the uncovering, denial, and final acceptance of the axiom of choice.

2.26   **(Project)** Prepare a historical account of the continuum hypothesis.

3

Metric Properties

In this chapter we meld our synthetic and analytic approaches with geometric notions. Geometric intuition can lead us to certain results with great ease. We investigate the notions of metrics, open and closed sets, intervals, connectedness, and compactness.

3.1 The Real Line

We are all hardwired with a Platonic notion of an ideal line—straight, infinitely divisible, and infinite in extent. Visualize such a line. Choose a unit length, then with a compass mark off points at this unit distance apart. Assign 0 to one of these marks, then continue to label the remaining marks with the increasing positive integers to the right, negative to the left as in figure 3.1.

Next, for each $n \in \mathbf{N}$ refine the integral marks of figure 3.1 by dividing each interval between integers into n equal parts and label with the appropriate rationals n'/n, $n' \in \mathbf{Z}$, as in figure 3.2.

Intuitively, these rational marks are "dense." Each point p on the line is singled out by the set of rational marks to its left or the rational marks to its right; the point p appears to capture geometrically the notion of a supremum and infimum. Let us leap to the belief that $\mathbf{R}$ can be mapped bijectively to points on this line so that "$x < y$"

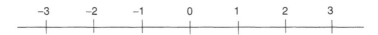

Figure 3.1 A line is marked off at equal distances and labeled with integers.

Figure 3.2 Line segments between integral marks are divided into n equal segments and marked with the corresponding rational number $(mn + k)/n$.

translates to "x is to the left of y." Henceforth we will visualize $\mathbf{R}$ as this *real line*.

3.2 Distance

The *absolute value* of $x \in \mathbf{R}$ is defined to be the nonnegative real

$$|x| = \begin{cases} x & \text{if } x > 0 \\ 0 & \text{if } x = 0 \\ -x & \text{if } x < 0. \end{cases} \tag{3.1}$$

Theorem A. The following equalities hold for all $x, y \in \mathbf{R}$:

$$|x \cdot y| = |x| \cdot |y|, \tag{3.2a}$$

$$\left|\frac{x}{y}\right| = \frac{|x|}{|y|} \quad \text{when } y \neq 0, \tag{3.2b}$$

$$x^2 = |x|^2, \quad \sqrt{x^2} = |x|, \tag{3.2c}$$

and the important *triangle inequality* $|x + y| \leq |x| + |y|$. $\tag{3.2d}$

Proof. The statements (3.2a)–(3.2c) are left as exercise 3.1. As for (3.2d), since $2xy \leq 2|x| \cdot |y|$, adding $x^2 + y^2$ to both sides yields $x^2 + 2xy + y^2 \leq x^2 + 2|x| \cdot |y| + y^2 = |x|^2 + 2|x| \cdot |y| + |y|^2$, that is,

$$|x + y|^2 \leq (|x| + |y|)^2. \tag{3.3}$$

Taking square roots (exercise 3.2) yields the result.

Definition. In concord with our geometric intuition, the *distance* between the real numbers x and y is

$$d(x, y) = |y - x|. \tag{3.4}$$

Theorem B. The distance metric $d(\cdot, \cdot)$ satisfies three fundamental properties: For all $x, y, z \in \mathbf{R}$,

I. $d(x, y) \geq 0$, with equality when and only when $x = y$.
$\tag{3.5a}$

II. $d(x, y) = d(y, x)$. $\tag{3.5b}$

III. $d(x, z) \leq d(x, y) + d(y, z)$. $\tag{3.5c}$

Proof. Exercise 3.4.

Properties I–III capture the concept of distance in Euclidean geometries of any dimension. Property III is called the *triangle inequality*, for in the Euclidean plane $\mathbf{R} \times \mathbf{R}$, when x, y, z denote the vertices of a triangle, property III reflects the fact that the length of one side is at most the sum of the lengths of the other two sides.

3.3 Topology

Fix a point x_0. The set of all points

$$B = \{x; \ d(x, x_0) < \epsilon\} \tag{3.6}$$

of distance from x_0 of less than ϵ is called the *open ball (neighborhood) of radius ϵ centered at x_0*.

For the real line this means B consists simply of all points x that satisfy $x_0 - \epsilon < x < x_0 + \epsilon$; that is, B is the *open interval $B = (x_0 - \epsilon, x_0 + \epsilon)$*. See figure 3.3.

Definition. A set $S \subset \mathbf{R}$ is *open* if every point $x_0 \in S$ possesses an open ball centered at x_0 wholly contained within S.

Note that the empty set $\emptyset$ is open by logical default.

Example 1. Each of the following intervals are open:

$$(a, b) = \{x; \ a < x < b\},$$
$$(a, \infty) = \{x; \ a < x\},$$
$$(-\infty, a) = \{x; \ x < a\},$$
$$(-\infty, \infty) = \mathbf{R}.$$

Theorem C. The collection Ω of open subsets of $\mathbf{R}$ possesses three fundamental properties:

    I. Both $\emptyset$ and $\mathbf{R}$ are open,

    II. the union of open sets is open, and

    III. the intersection of a finite number of open sets is open.

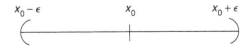

Figure 3.3 An open ball in $\mathbf{R}$ is simply an open interval centered at x_0.

Proof. Exercise 3.5. Also see exercise 3.28.

A set $S \subset \mathbf{R}$ is *closed* if its complement is open.

Example 2. Each of the following intervals are closed:

$$[a, b] = \{x; \ a \le x \le b\},$$
$$[a, \infty) = \{x; \ x \ge a\},$$
$$(-\infty, a] = \{x; \ x \le a\},$$
$$(\infty, \infty) = \mathbf{R}.$$

The only sets that are both open and closed at the same time are $\emptyset$ and $\mathbf{R}$ (exercise 3.6). Intuitively, most sets are neither open nor closed. For example, delete any one point from the closed interval $[a, b]$, $a < b$, to obtain a set that is neither open nor closed. A concise means of describing open and closed sets is via their boundaries.

Definition. The point x_0 is a *boundary point* of the set S if every open neighborhood of x_0 meets both S and its complement S'. For example, the intervals (a, b) and $[a, b]$ possess the same boundary points a and b.

Corollary. A set is open exactly when it contains *none* of its boundary points. A set is closed exactly when it contains *all* of its boundary points.

Proof. Exercise 3.8.

Example 3. Every real number is a boundary point of the set of rational numbers $\mathbf{Q}$ (exercise 3.15).

Example 4. The set $S = \{1, 1/2, 1/3, \dots, 1/n, \dots\}$ has each of its elements $1/n$ as boundary points, since every open neighborhood $(1/n - \epsilon, 1/n + \epsilon)$ contains points of S, (namely $1/n$) as well as points not in S. Moreover, 0 is also a boundary point of S, since every open neighborhood about 0 contains points in S (namely $1/n$ for all sufficiently large n) and points not in S (namely 0 itself). Thus in notation now in fashion for the boundary points of a set, $\partial S = S \cup \{0\}$.

3.4 Connectedness

A set S is *connected* if it not disconnected. A set S is *disconnected* if there exist two disjoint open sets A and B that both meet S and together cover S, that is, $A \cap B = \emptyset$, $A \cap S \ne \emptyset$, $B \cap S \ne \emptyset$, and $S \subset A \cup B$.

Theorem D. The connected subsets I of **R** are exactly the *intervals,* that is, subsets I with the property

$$a, b \in I \quad \text{implies} \quad [a, b] \subset I. \tag{3.7}$$

Proof. Suppose I is connected and $a, b \in I$. Let $a < c < b$. Set $A = (-\infty, c)$ and $B = (c, \infty)$. The open intervals A and B will disconnect I unless $c \in I$. Thus I is an interval.

Conversely, suppose I is an interval, yet there exists a disconnection by open sets A and B. Interchanging the letters A and B if necessary, choose $a \in A \cap I$ and $b \in B \cap I$ with $a < b$. Thus $[a, b] \subset I$. Consider the supremum c of all $x \in [a, b] \cap A$. Since $c \in I$, either $c \in A$ or $c \in B$. But all balls of small radii about c must eventually lie wholly within either A or B, respectively. Moreover, all such small intervals $(c - \delta, c + \delta)$ must contain points in A left of c and points of B to its right. But A and B cannot meet. Thus I is connected.

3.5 Compactness

A subset K of **R** is said to be *compact* if every open covering of K has a finite subcovering. That is, if

$$K \subset \bigcup_{\alpha} O_{\alpha},$$

where the O_{α} are open, then almost all of these O_{α} are unnecessary, that is, there are finitely many subscripts $\alpha_1, \alpha_2, \ldots, \alpha_n$ so that

$$K \subset \bigcup_{k=1}^{n} O_{\alpha_k}.$$

Example 5. The open interval $I = (0, 1)$ has the infinite open covering

$$I \subset \bigcup_{k=1}^{\infty} (0, 1 - \frac{1}{k}) \tag{3.8}$$

but no finite number of the intervals $(0, 1 - 1/k)$ will suffice (exercise 3.17). See figure 3.4. Thus $I = (0, 1)$ is not a compact set. In contrast, any finite set is clearly compact.

It is now time to tackle a significant result, a foundation of calculus, and the most difficult proof of this book. The rest of calculus is an easy downhill slide from here.[1]

---

[1]One of my teachers, Paul Halmos, postulates three "laws" of mathematics. His third is the *conservation of work*: Results obtained easily cannot be important. As you climb

Figure 3.4 The open interval (0, 1) is covered by infinitely many open intervals (0, 1 − 1/k). No finite number of these intervals will suffice.

Theorem E. The compact sets of **R** are exactly the sets that are both closed and bounded.

Proof. Assume K is compact. Note that K (like any subset of **R**) has the open covering

$$K \subset \bigcup_{k=1}^{\infty} (-k, k). \tag{3.9}$$

Since a finite subcover will suffice, for some large n,

$$K \subset \bigcup_{k=1}^{n} (-k, k), \tag{3.10}$$

that is, $K \subset (-n, n)$ and so K is bounded. It remains to show K is closed.

For the sake of an eventual contradiction, suppose x_0 is a boundary point of K not in K. For each point $x \in K$, choose an open ball B_x centered at x and an open ball P_x centered at x_0 that do not meet. Thus we have the covering

$$K \subset \bigcup_{x \in K} B_x, \tag{3.11}$$

which by compactness must possess the finite subcover

$$K \subset \bigcup_{k=1}^{n} B_{x_k}. \tag{3.12}$$

---

to each higher plateau of knowledge, no matter by what path, you must perform a pre-determined minimum amount of work needed to reach that plateau. (All graders know this intuitively—if a student proposes an easy proof of a hard result, the proof is flawed.)

But then

$$P_0 = \bigcap_{k=1}^{n} P_{x_k} \tag{3.13}$$

is the intersection of a finite number of open sets and is hence an open set containing x_0 that does not meet K. Thus x_0 cannot be a boundary point of K. We have shown K compact implies K is both closed and bounded.

Conversely, assume K is closed and bounded. Our goal is to prove K compact. That is, in any open covering

$$K \subset \bigcup_{\alpha} O_\alpha, \tag{3.14}$$

all but a finite number of the open sets O_α may be discarded.

Consider any one of the open sets O_α of the covering (3.14). Each point $x \in O_\alpha$ is in an open interval $I_{\alpha\beta}$ that is centered at a rational with rational radius and that lies wholly within O_α. We may enlarge the covering (3.14) by replacing each O_α by all of these intervals $I_{\alpha\beta}$ it contains to obtain the covering

$$K \subset \bigcup_{\alpha,\beta} I_{\alpha\beta}. \tag{3.15}$$

But *the number of open balls with rational radii centered at rationals is countable* (exercise 3.18), and so after resubscripting the intervals $I_{\alpha,\beta}$ the covering (3.15) becomes the covering

$$K \subset \bigcup_{n=1}^{\infty} I_n.$$

For each n (using the axiom of choice) select one $O_\alpha \supset I_n$ denoted by O_n to obtain the countable subcover

$$K \subset \bigcup_{k=1}^{\infty} O_k. \tag{3.16}$$

This is significant progress—we have proved the *Lindelöf principle: Any infinite open cover has a countable subcover.*

Suppose no finite subcover of (3.16) exists. Then for each natural number n, there exists a point $x_n \in K$ yet

$$x_n \notin \bigcup_{k=1}^{n} O_k. \tag{3.17}$$

Let X be the set of all distinct values among the x_n. Because of (3.17), X is an infinite set.

Since K is bounded and $X \subset K$, it follows from the Heine-Borel lemma below that there must exist at least one *point of accumulation* $c \in X$; that is, every open neighborhood of c contains infinitely many points of X. Since K is closed, $c \in K$. But then some element O_N of the cover must contain c, i.e., $c \in O_N$, and hence O_N must contain infinitely many points x_n of X, an impossibility because of (3.17). Thus a finite subcover must exist; K is compact.

Heine-Borel lemma. An infinite bounded set possesses at least one accumulation point.

Proof.[2] Suppose X is an infinite but bounded set. Find bounds a, b so that $X \subset [a, b]$. Let $m = (a+b)/2$. Because X is infinite, either the left-half interval $[a, m]$ or the right-half interval $[m, b]$ contains infinitely many points of X. Choose one of these half-intervals that meets X infinitely often and label its left endpoint a_1 and its right endpoint b_1. Let $m_1 = (a_1 + b_1)/2$. Thus either $[a_1, m_1]$ or $[m_1, b_1]$ meets X infinitely often, and so forth.

In this manner we obtain a nondecreasing sequence of the left endpoints a_k bounded above by each term of the nonincreasing sequence of right endpoints b_k of intervals $[a_k, b_k]$ meeting X infinitely often, that is,

$$a \leq a_1 \leq a_2 \leq a_3 \leq \cdots\cdots \leq b_3 \leq b_2 \leq b_1 \leq b. \tag{3.18}$$

Exactly one number is trapped between these two sequences, namely

$$c = \sup a_n. \tag{3.19}$$

This supremum c is a point of accumulation (exercise 3.19).

---

[2]What follows is the *method of bisection*, which we will employ again as a root-finding algorithm in the chapter 4.

Exercises

3.1   Verify (3.2a)–(3.2c).

3.2   Prove for any positive reals $0 < x < y$ that $0 < x^{1/N} < y^{1/N}$ for $N \in \mathbf{N}$.
       Hint: Assume the conclusion false and apply exercise 2.21.

3.3   Describe how to construct the rational m/n with a compass and a
       straightedge.

3.4   Prove the three metric properties of theorem B.

3.5   Prove the fundamental three properties of open sets given in
       theorem C.

3.6   Prove that only subsets of $\mathbf{R}$ that are are both open and closed are $\varnothing$
       and $\mathbf{R}$.
       Hint: Such a set A and its complement B would disconnect $\mathbf{R}$.

3.7   Delete one point from a closed interval $[a, b]$ with $a < b$. Show that
       the resulting set is neither open nor closed.

3.8   Prove that a set is open exactly when it contains none of its boundary
       points and closed exactly when it contains all of its boundary points.

3.9   Suppose for a set X there is a distance function $d : X \times X \longrightarrow \mathbf{R}$ with
       the three properties of (3.5a)–(3.5c). Then the pair (X, d) is called a
       metric space.
            For example, consider the silly distance function $d^{\#}(x_1, x_2) = 1$ if
       $x_1 \neq x_2$, 0 otherwise. Show that $(X, d^{\#})$ is a metric space. Which
       subsets of X are open?

3.10   Show that $d_{1/2}(x, y) = \sqrt{|x - y|}$ also satisfies the three fundamental
       metric properties (3.5). What are the resulting open sets?

3.11   Let f be any increasing function on the set of nonnegative reals with
       the properties $f(0) = 0$ and $f(x + y) \leq f(x) + f(y)$ for all $x, y \geq 0$. Show
       $d_f(x, y) = f(|x - y|)$ is a metric on $\mathbf{R}$.

3.12   Show that

$$d_0(x, y) = \frac{|x - y|}{1 + |x - y|}$$

is a metric on **R** with the identical open sets as the usual metric, yet no two points are further than 1 unit apart; the open ball about 0 of radius 1 includes all real numbers!

3.13   Invent a truly bizarre metric on **R**. Which sets are open with respect to your metric?
        Suggestion: Prove that when $f : \mathbf{R} \to \mathbf{R}$ is injective, then $d(x, y) = |f(x) - f(y)|$ is a metric on **R**. Now make strange choices for f.

3.14   Show that **Q** has every point of **R** as a boundary point.

3.15   Find a set S with exactly seven boundary points.

3.16   Discover all nine interval types I, that is, subsets of **R** satisfying (3.7).
        Partial answer: $[a, b]$, $[a, b)$, (a, ∞), $\ldots$.

3.17   Show that the infinite cover (3.8) cannot be reduced to a finite subcover.

3.18   Prove that the collection of all intervals of rational radii centered at rationals is a countable. Prove the more general result, *a countable union of countable sets is countable.*
        Hint: Use Cantor's diagonalization of exercise 1.25.

3.19   Verify that the c of (3.19) is an accumulation point of X in the proof of the Heine-Borel lemma.

3.20   Let $X \subset \mathbf{R}$. A point $x \in X$ is an *interior* point of X if it possesses an open neighborhood wholly contained by X. The set of all interior points of X is often denoted by X°. A point y is *exterior* to X if it is an interior point of the complement X'.
        Prove that **R** is the disjoint union $\mathbf{R} = X^\circ \cup \partial X \cup X'^\circ$. (The symbol ∂X is the widely used notation for the set of boundary points of X.)

3.21   Give an example of a set with no points of accumulation yet with infinitely many boundary points.

3.22   Give an example of a set with no boundary points yet with infinitely many accumulation points.

3.23   The *closure* of a set X is the set $\bar{X} = X \cup \partial X$. Prove that $\bar{X}$ is the intersection of all closed sets containing X.

3.24   Prove that the intersection of a decreasing sequence of nonempty compact sets is nonempty.

3.25   What is the maximum number of distinct sets that can be generated from any one set by repeated operations of closure and complementation?
       Answer: 14.

3.26   Show by an example that the intersection of an infinite number of open sets may not be open.

3.27   Prove that a closed subset of a compact set is itself compact.

3.28   Suppose X is a set and Ω a family of subsets of X (to be called the *open* subsets) with three properties:

       I. $\emptyset, X \in \Omega$,
      II. Ω is closed under unions, and
      III. Ω is closed under finite intersections.

Then the pair (X, Ω) is called a *topological space*.
       Establish that every set X can be made into an uninteresting topological space by either declaring all sets open (the *discrete* topology), or by declaring that only $\emptyset$ and X are open (the *indiscrete* topology). Show that in contrast to the discrete topology, the indiscrete topology cannot arise from a metric.

3.29   Prove that every open set of **R** is the disjoint union of a finite or countably infinite collection of open intervals.
       Outline: Every set is the disjoint union of its *components*, its maximal connected subsets. But the components of an open set are open. Apply the Lindelöf principle.

4

Continuity

In this chapter we investigate the important property of continuity and its consequences—that the composition of continuous functions is continuous and that the continuous images of connected or compact sets are again connected or compact, respectively, thus obtaining the first two pillars of calculus: the intermediate value theorem and the theorem on the achievement by continuous functions of their extreme values. We will finish with the numerical root-finding method of bisection.

4.1 The Definition

A function $f : \mathbf{R} \longrightarrow \mathbf{R}$ is said to be *(everywhere) continuous* if the inverse image $f^{-1}(V)$ under f of every open set V is again an open set. See figure 4.1.

Example 1. The function $f(x) = x$ is clearly continuous, since the preimage of every open set V is the open set V. Or, more generally, the *linear* function $f(x) = mx + b$ is everywhere continuous. To see this there are, however, two cases. If $m = 0$ then f is constant and hence continuous by the argument of example 2. If $m \neq 0$, then the preimage of an open set V under this linear function is (exercise 4.1) the open set $U = (V - b)/m$.

Example 2. Constant functions $f(x) = c$ are continuous. For let V be any open set. Then $f^{-1}(V)$ is either $\mathbf{R}$ (if $c \in V$) or $\emptyset$ (if $c \notin V$).

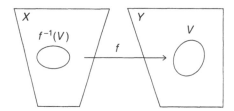

Figure 4.1 A function f is continuous from the metric space X into the metric space Y if the preimage of every open set V of Y under f is an open set of X.

4.2 Consequences

Theorem A. *The composition of two continuous functions is again continuous.*

Proof. Let f and g be continuous and set $h = g \circ f$; that is, $h(x) = g(f(x))$. Let V be open. Then $h^{-1}(V) = f^{-1}(g^{-1}(V))$. Since $U = g^{-1}(V)$ is open, so is $f^{-1}(U)$. See figure 4.2.

Example 3. The continuity of $h(x) = \text{Arctan} \sin e^{x^2}$ follows from the continuity (to be established) of the rules $x \mapsto \text{Arctan}\, x$, $\sin x$, e^x, x^2 that are factors of the composition h.

Theorem B. (the intermediate value theorem) *The continuous image of a connected set is connected.* For maps from $\mathbf{R}$ to $\mathbf{R}$ this means: *Continuous functions map intervals to intervals.*

Proof. Suppose C is a connected set and f is continuous. Let $D = f(C)$. Suppose D is disconnected; that is, there exist open sets A and B that disconnect D. That is,

$$A \cap B = \emptyset, \ A \cap D \neq \emptyset, \ B \cap D \neq \emptyset, \ D \subset A \cup B. \tag{4.1}$$

But then $f^{-1}(A)$ and $f^{-1}(B)$ are open sets disconnecting C (exercise 4.2), a contradiction. Thus D is connected.

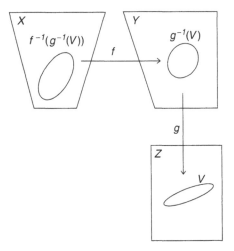

Figure 4.2 The composition of two continuous functions is continuous. The inverse image of the inverse image of an open set is open.

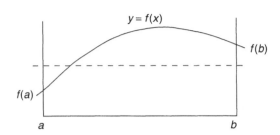

Figure 4.3 A continuous function maps intervals
to intervals. The image of $[a, b]$ under f includes
at the very least all values between $f(a)$ and $f(b)$.

Informal Version. The intermediate value theorem (IVT) states that if
a continuous function $f : \mathbf{R} \longrightarrow \mathbf{R}$ takes on the values $c = f(a)$ and
$d = f(b)$, then as x runs from a to b, f takes on at least all of the values
between c and d. Geometrically, the graph of f crosses (at least once)
every horizontal line $y = y_0$ between the lines $y = c$ and $y = d$. See
figure 4.3.

Example 4. As we shall soon prove, all polynomials are continuous. So
consider the cubic $f(x) = x^3 + x - 1$. Note that $f(0) = -1$ while $f(1) = 1$.
Thus by the IVT, f possesses a *zero*; that is, $f(x) = 0$ possesses a *root*,
somewhere between $x = 0$ and $x = 1$. We will exploit this observation
to write a root-finding algorithm in §4.4.

Theorem C. *The continuous image of a compact set is compact.*

Proof. Suppose C is compact and f is continuous. Let $K = f(C)$. Any
open covering

$$K \subset \bigcup_{\alpha} O_{\alpha} \tag{4.2}$$

yields the open covering

$$C \subset \bigcup_{\alpha} f^{-1}(O_{\alpha}), \tag{4.3}$$

which by the compactness of C possesses a finite subcover

$$C \subset \bigcup_{k=1}^{n} f^{-1}(O_{\alpha_k}). \tag{4.4}$$

The subcover (4.4), when mapped back through f, forms a finite subcover

$$K \subset \bigcup_{k=1}^{n} O_{\alpha_k}. \tag{4.5}$$

Therefore K is compact.

Corollary. A continuous function f on a closed and bounded interval $I = [a, b]$ is not only bounded, but it achieves its maximum and minimum values.

Proof. Since $I = [a, b]$ is closed and bounded, it is compact. Thus by the theorem $K = f(I)$ is compact, hence closed and bounded (theorem E, §3.5). By the IVT, K is an interval, and so $K = [c, d]$. Thus the maximum of f on I is d and its minimum is c.

4.3 Combinations of Continuous Functions

We may combine continuous functions in many different ways to obtain continuous functions of ever increasing complexity. We have already seen in theorem A that the composition of continuous functions is again continuous.

Theorem D. *The sum of continuous functions is continuous.*

Proof. Suppose f and g are continuous. We will obtain the sum function $h = f + g$, given by the rule $h(x) = f(x) + g(x)$, as a composition of two mappings: namely $i : \mathbf{R} \longrightarrow \mathbf{R} \times \mathbf{R}$ given by

$$i(x) = (f(x), g(x)) \tag{4.6}$$

followed by the addition map $a : \mathbf{R} \times \mathbf{R} \longrightarrow \mathbf{R}$ given by

$$a(y, z) = y + z. \tag{4.7}$$

After proving that each map is continuous, we will have our result by theorem A.

First, consider the map of (4.6). Let V be any open set in $\mathbf{R}^2 = \mathbf{R} \times \mathbf{R}$ under the usual distance metric

$$d(p_1, p_2) = |p_1 - p_2| = \sqrt{(y_1 - y_2)^2 + (z_1 - z_2)^2}, \tag{4.8}$$

and let U be its inverse image under the map i, i.e., $U = i^{-1}(V)$. We must prove that U is open.

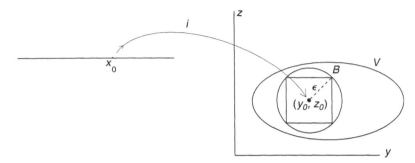

Figure 4.4 The map $i(x) = (f(x), g(x))$.

Let $x_0 \in U$ and set $i(x_0) = (f(x_0), g(x_0)) = (y_0, z_0)$. Find an open ball B of radius ϵ about $p = (y_0, z_0)$ wholly contained by V. Note for its inscribed square of half-side length $\epsilon/\sqrt{2}$,

$$(y_0 - \epsilon/\sqrt{2}, y_0 + \epsilon/\sqrt{2}) \times (z_0 - \epsilon/\sqrt{2}, z_0 + \epsilon/\sqrt{2}) \subset B \subset V. \qquad (4.9)$$

See figure 4.4.

But then because both f and g are continuous, the set

$$O = f^{-1}((y_0 - \epsilon/\sqrt{2}, y_0 + \epsilon/\sqrt{2})) \cap g^{-1}((z_0 - \epsilon/\sqrt{2}, z_0 + \epsilon/\sqrt{2})) \qquad (4.10)$$

is an open neighborhood of x_0 with $i(O) \subset B \subset V$ and hence $O \subset U$. Thus the map i of (4.6) is continuous.

It remains to show the addition map a of (4.7) is continuous. Let W be any open subset of $\mathbf{R}$ and let $V = a^{-1}(W)$. We must prove V open. Pick any point $p = (y_0, z_0) \in V$ and set $w_0 = a(y_0, z_0) = y_0 + z_0$. Choose $\epsilon > 0$ small enough so that the ball $(w_0 - \epsilon, w_0 + \epsilon) \subset W$. Consider a new ball B of radius $\delta = \epsilon/2$ about $p = (y_0, z_0)$. Then for any $(y, z) \in B$, certainly $|y - y_0| < \epsilon/2$ and $|z - z_0| < \epsilon/2$, and so[1]

$$|a(y, z) - w_0| = |y + z - y_0 - z_0| \leq |y - y_0| + |z - z_0| < \frac{\epsilon}{2} + \frac{\epsilon}{2} = \epsilon, \qquad (4.11)$$

and so $a(B) \subset W$, proving that $a^{-1}(W)$ is open and hence addition is continuous.

Theorem E. *The product of continuous functions is continuous.*

---

[1] Experienced mathematicians dismiss this as "merely an epsilon over two" proof.

Proof. Suppose f and g are continuous. We will obtain the product function $h = fg$, given by the rule $h(x) = f(x) \cdot g(x)$, as a composition of two mappings, namely $i : \mathbf{R} \longrightarrow \mathbf{R} \times \mathbf{R}$ given by (4.6) followed by the multiplication map $m : \mathbf{R} \times \mathbf{R} \longrightarrow \mathbf{R}$ given by

$$m(x, y) = x \cdot y. \tag{4.12}$$

We have already shown in the proof of theorem D that i is continuous. It remains to show multiplication (4.12) is continuous.

Let $W \subset \mathbf{R}$ be open and let $V = m^{-1}(W)$. Choose $p = (y_0, z_0) \in V$ and set $w_0 = m(y_0, z_0) = y_0 \cdot z_0$. Choose $\epsilon > 0$ small enough so that $(w_0 - \epsilon, w_0 + \epsilon) \subset W$. Slelect any positive $\delta < 1$ with

$$\delta < \frac{\epsilon}{1 + |y_0| + |z_0|}. \tag{4.13}$$

Then for any point (y, z) in the open ball B about (y_0, z_0) of radius δ, certainly $|y - y_0| < \delta$ and $|z - z_0| < \delta$. But then we have[2]

$$
\begin{aligned}
|m(y, z) - w_0| &= |yz - y_0 z_0| = |yz - y_0 z + y_0 z - y_0 z_0| \\
&\leq |yz - y_0 z| + |y_0 z - y_0 z_0| = |y - y_0| \cdot |z| + |y_0| \cdot |z - z_0| \\
&\leq \delta|z| + |y_0|\delta < \delta(1 + |z_0|) + |y_0|\delta = \epsilon.
\end{aligned} \tag{4.14}
$$

Thus $m(B) \subset W$, which proves multiplication is continuous.

Corollary. *Polynomials are everywhere continuous.*

Proof. As we established above in example 1, the function given by $f(x) = x$ is clearly continuous. But then so is its square, its cube, and so on. That is, $g(x) = x^n$ is continuous for any natural number exponent n. But in example 2 we showed constant functions are continuous. Thus any rule of the form $f(x) = cx^n$ is continuous. Finally, the sum of any such functions must again be continuous; that is, any *polynomial function*

$$f(x) = c_0 + c_1 x + c_2 x^2 + \cdots + c_n x^n \tag{4.15}$$

is continuous.

Synopsis of the ϵ–δ proof method. Let us abstract the method that we repeatedly employed to establish theorems D and E. Our mission was to prove the continuity of several maps $f : X \longrightarrow Y$, where X and Y were two metric spaces. That is, given any open set $V \subset Y$, and $U = f^{-1}(V)$, we were required to prove that U is open in X.

---

[2]Here we employ *trick #1 of mathematics:* We add and subtract the same quantity.

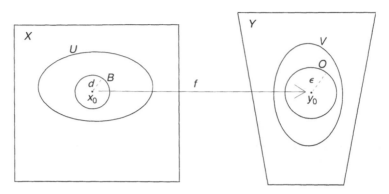

Figure 4.5 The $\epsilon-\delta$ proof method: For each ϵ neighborhood of the image point $y_0 = f(x_0)$, find a δ neighborhood of the point x_0 that is mapped by f into the ϵ neighborhood.

To accomplish this, we selected an arbitrary point $x_0 \in U$. Our goal, then, is reduced to finding some open ball B about x_0 that is wholly contained within U.

To this end, we chose a sufficiently small positive ϵ so that the open ball O of radius ϵ about $y_0 = f(x_0)$ is contained wholly within V. Then, after doing scratchwork offline, we obtained a positive number δ arising from ϵ and x_0 so that the entire ball B of radius δ centered at x_0 is mapped by f into $O \subset V$, giving that $B \subset U$. But since $x_0 \in U$ was chosen arbitrarily, U is open. See figure 4.5.

In short, we proved at each $x_0 \in X$, *for every $\epsilon > 0$ there exists a $\delta > 0$ such that for all $x \in X$,*

$$d_X(x, x_0) < \delta \quad \text{implies} \quad d_Y(f(x), f(x_0)) < \epsilon \tag{4.16}$$

(where d_X, d_Y are the metrics on X and Y, respectively).

This proof scheme is the central proof scheme of calculus. It is refered to as an "epsilon and delta proof." The exercises provide experience with this proof technique.

4.4 Bisection

One obvious consequence of the intermediate value theorem is: *If a continuous function $f : \mathbf{R} \longrightarrow \mathbf{R}$ changes sign from one point to another, then it must possess a zero between.* This simple observation yields a slow but sure computer root-finding algorithm.

Bisection Algorithm. Assumptions: f continuous and $f(a) \cdot f(b) < 0$ where $a < b$.

Pseudocode:

```
loop
    m = (a + b)/2
    if f(a) · f(m) < 0,  then b = m
        else a = m
until
    b − a < required accuracy.
```

At each iteration of the loop, the location of the zero is narrowed to half its previous indeterminacy.

Example 5. Let us estimate a root of $x^3 + x - 1 = 0$ that must lie somewhere within the interval $I = [0, 1]$. The existence of such a root is guaranteed, since $f(x) = x^3 + x - 1$ has the sign change $f(0) = -1$ and $f(1) = 1$. There is a zero somewhere in $[0, 1]$.

First iteration. $m = (1 + 0)/2 = 0.5$. Note that

$$f(m) = f(0.5) = \left(\frac{1}{2}\right)^3 + \frac{1}{2} - 1 = -0.375 < 0,$$

the same sign as $f(0) = -1$. Our zero lies in $[0.5, 1]$.

Second iteration. $m = (0.5 + 1)/2 = 0.75$. Note that $f(0.75) = 0.171875$, a change of sign from $f(0.5)$. Our zero lies in $[0.5, 0.75]$.

Third iteration. $m = (0.5 + 0.75)/2 = 0.625$. Note that $f(0.625) = -0.13085$, no change of sign from $f(0.5)$. Our zero lies in $[0.625, 0.75]$.

Fourth iteration. $m = (0.625 + 0.75)/2 = 0.6875$. Note that $f(0.6875) = 0.01245\cdots$, a change of sign from $f(0.625)$. Our zero lies in $[0.625, 0.6875]$.
And so on. This algorithm is very slow but has no fatal flaws—it will always converge to the correct value for the zero. See the exercises for practice with this algorithm.

4.5 Subspace Topology
We now extend the notion of continuity to functions such as $\ln x, \tan x$, or $\sqrt{x(1-x)}$ that are not defined everywhere on **R**. The definition is essentially the same as before: A function $f : X \longrightarrow \mathbf{R}$ is *continuous on the domain X* if the inverse image of every open set $V \subset \mathbf{R}$ under f is open *relative to X*.

A subset $P \subset X$ is *open relative to X* if $P = U \cap X$ for some open set $U \subset \mathbf{R}$. Put another way, the relatively open sets are open with respect to the ordinary metric cut back to only the points of X.

Example 6. If $X = [0, 1]$, then $P = (0, 1]$ is open relative to X, since $P = X \cap (0, 2)$. Or, for another example, if $X = \mathbf{Q}$, then the relative open sets are the sets of rationals contained in open subsets of $\mathbf{R}$.

Remark. We have been careful in the statements of previous definitions and results—all results and proofs remain correct when "open," "closed," "connected," "compact" are replaced by "relatively open," "relatively closed," and so on. See exercises 4.15–4.18.

Example 7. Let us prove $f(x) = 1/x$ is continuous on $X = (0, \infty)$. Fix $x_0 > 0$ and let $\epsilon > 0$ be given. Let $\delta = \min(x_0/2, \epsilon x_0^2/2)$. Assume $|x - x_0| < \delta$. Then $x > x_0/2 > 0$ and

$$|f(x) - f(x_0)| = |\frac{1}{x} - \frac{1}{x_0}| = \frac{|x - x_0|}{x \cdot x_0} < \frac{\delta}{(x_0/2) \cdot x_0} \le \epsilon. \qquad (4.17)$$

Thus the inverse image of every open set V under f is relatively open in $X = (0, \infty)$.

Exercises

4.1   Prove that if V is an open set, then so is its *translate* $V - b = \{x; x = v - b, v \in V\}$ as well as its stretch $aV = \{x; x = av, v \in V\}$ provided $a \neq 0$.

4.2   Finish the proof of the intermediate value theorem B.

4.3   If f is continuous, prove that the inverse image of every closed set K under f is again closed. Prove that the converse also holds.

4.4   Deduce from the IVT that there are two diametrically opposite points on a circle of radius fifty miles from your present position that are at exactly the same temperature.
      Hint: Consider the function $f(\theta) = T(\theta + \pi) - T(\theta)$, where $T(\theta)$ is the temperature at compass direction θ.

4.5   Show by example that the continuous image of an open set need not be open.
      Suggestion: Use $f(x) = x^2$.

4.6   Show by example that the continuous image of a closed set need not
        be closed.

4.7   To obtain practice with the $\epsilon-\delta$ proof method, prove directly that
        $f(x) = 2x - 3$ is continuous without recourse to theorems D or E.
            Outline: Let $\epsilon > 0$ be given. Let $\delta = \epsilon/2$ and assume $|x - x_0| < \delta$.
        Then $|f(x) - f(x_0)| = \cdots < 2\delta = \epsilon$.

4.8   Prove directly via an $\epsilon-\delta$ proof method that $f(x) = x^2$ is continuous.
            Outline: Let $\epsilon > 0$ be given. Choose $\delta = \min(1, \epsilon/(1 + 2|x_0|)$ Assume
        $|x - x_0| < \delta$. Then $|f(x) - f(x_0)| = \cdots < (1 + 2|x_0|)\delta \leq \epsilon$.

4.9   Prove directly via an $\epsilon-\delta$ proof method that $f(x) = x^3$ is continuous.

4.10   Prove directly via an $\epsilon-\delta$ proof method that $f(x) = 1/(1 + x^2)$ is
        continuous.

4.11   Prove that *continuous functions vanish on closed sets*. Show that the
        points where a continuous function is positive form an open set.
        Deduce that two continuous functions agree on a closed set.

4.12   Perform three iterations of bisection to estimate the zero of
        $f(x) = x^2 - 5$ that lies in $[2, 3]$.

4.13   Perform four iterations of bisection to estimate the root of $x = \cos x$
        (in radian measurement) on $[0, 1]$.

4.14   **(Project)** The tracking of satellites requires solving *Kepler's equation*,
        $M = E - e \sin E$, where M is the *mean anomaly*, E the *eccentric anomaly*,
        and e the *eccentricity* of the elliptical orbit [Moulton]. How many
        iterations of bisection are necessary to solve $1 = E - 0.1 \sin E$ for the
        eccentric anomaly E to three decimal places?

4.15   Prove that a relatively open subset P of an open set X is open.

4.16   Prove that a relatively closed subset C of a closed set X is closed.

4.17   Prove that a relatively connected set is connected. Does the converse
        hold?

4.18   Prove that a relatively compact set is compact and conversely.

4.19   Prove via $\epsilon-\delta$ that $f(x) = 1/x^2$ is continuous on $X = (0, \infty)$.

4.20   Prove via $\epsilon-\delta$ that $f(x) = \sqrt{x(1-x)}$ is continuous on $X = [0, 1]$.

4.21   Consider *Dirichlet's example:* $f(x) = 1$ when $x \in \mathbf{Q}$, 0 otherwise. Prove f is nowhere continuous.

4.22   Consider the *ruler function:* For the nonzero rational $x = a/b$ in lowest terms with $b > 0$, set $f(a/b) = 1/b$, 0 otherwise. Show f is continuous at every irrational but discontinuous at every nonzero rational.

4.23   Prove that an increasing function can possess at most a countable number of discontinuities.
          Hint: Apply the Lindelöf principle to the collection of all nonempty intervals $(f(x^-), f(x^+))$. (See exercise 5.33.)

4.24   Prove that the cartesian product of compact sets is compact.

4.25   Climbers begin at 6:00 a.m. and reach the summit by nightfall. The following morning they begin their descent at 6:00 a.m. Prove that at some time of day, they will be at the exact same altitude as the preceding day at that time.

4.26   Prove that the graph of a continuous function is a connected set (and thus "can be drawn without lifting the pencil").

4.27   We may rephrase the search for zeros of f as a search for *fixed points* of F (points x where $F(x) = x$) by setting $F(x) = f(x) + x$. Fixed points of F can occasionally be found by *iterating* F : Start with some initial guess $x = x_0$, then repeatedly apply F, i.e., $x_{n+1} = F(x_n)$. Does this method work for finding roots of $f(x) = x^2 - x$? See exercise 9.50.

5

Limits and Derivatives

In this chapter we examine the concept of a *limit,* in particular, the limit of the difference quotient known as the *derivative.* We discover the rules of differentiation and derive differentiation formulas for polynomial, inverse, and trigonometric functions.

5.1 Limits

Let us examine local behavior of functions.

Definition A. Suppose $f : X \subset \mathbf{R} \longrightarrow \mathbf{R}$ and that x_0 is either a boundary point or a point of the domain X of f. We say that L *is the limit of f as x approaches x_0* within X, in symbols

$$\lim_{x \to x_0} f(x) = L, \qquad (5.1)$$

when the preimage under f of each open set V containing L contains a punctured relatively open neighborhood P of x_0, i.e., a set of the form $P = U \cap (X \setminus \{x_0\})$, where U is an open set containing x_0. See figure 5.1.

This is equivalent (exercise 5.1) to the (in this case simpler) $\epsilon - \delta$ statement: For every $\epsilon > 0$ there is a $\delta > 0$ such that for all $x \in X$,

$$0 < |x - x_0| < \delta \quad \text{implies} \quad |f(x) - L| < \epsilon. \qquad (5.2)$$

Corollary A. If $x_0 \in X$ and f is continuous on X, then the limit of f at x_0 exists and equals $f(x_0)$. Conversely, if the limit of f exists at each

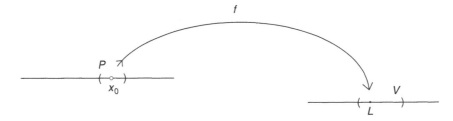

Figure 5.1 The limit of f is L as x approaches x_0 exactly when for every open neighborhood V of L, there is a punctured relatively open neighborhood P of x_0 that is mapped entirely by f into V.

point $x_0 \in X$ and equals $f(x_0)$, then f is continuous on X. In short, f is *continuous on X exactly when it is continuous at each point of X.*

Proof. Exercise 5.2.

Corollary B. If f possesses a limit at $x = x_0$ and g is everywhere continuous, then

$$\lim_{x \to x_0} g(f(x)) = g(\lim_{x \to x_0} f(x)). \tag{5.3}$$

Proof. Exercise 5.3.

Example 1. Consider the rule

$$f(x) = \frac{\text{Arctan } x}{x}, \tag{5.4}$$

with the nonzero reals as its domain X. What is its limit at $x_0 = 0$? Setting a calculator in radian measure, let us prepare a table of values of f near $x_0 = 0$ as in table 5.1. Even given the meager data of Table 5.1, we can leap to the conclusion that

$$\lim_{x \to 0} \frac{\text{Arctan } x}{x} = 1. \tag{5.5}$$

We will establish this analytically in exercise 5.24.

Example 2. At other times, simple algebra will reveal the limit. Consider the limit on the domain $X = \{x \in \mathbf{R}; \ x \neq 1\}$:

$$\lim_{x \to 1} \frac{x^2 - 3x + 2}{x - 1} = \lim_{x \to 1} \frac{(x-1)(x-2)}{x-1} = \lim_{x \to 1} (x - 2) = -1.$$

The cancellation of the common term $x - 1$ is valid, since $1 \notin X$.

TABLE 5.1
Values of the rule (5.4)
near $x_0 = 0$.

x	$f(x)$
0.1	0.996686525
−0.01	0.999966669
0.001	0.999999667

Theorem A. If two functions f and g on a common domain X both have limits at $x = x_0$ respectively, then so does their sum and product. In fact,[1]

$$\lim_{x \to x_0} (f(x) + g(x)) = \lim_{x \to x_0} f(x) + \lim_{x \to x_0} g(x), \qquad (5.6a)$$

$$\lim_{x \to x_0} (f(x) \cdot g(x)) = \lim_{x \to x_0} f(x) \cdot \lim_{x \to x_0} g(x). \qquad (5.6b)$$

Proof. Combining the continuity of the addition and multiplication maps (4.7) and (4.12) with (5.3) applied to $i(x) = (f(x), g(x))$ will yield the result (exercise 5.4). See §4.3.

Corollary A. A *rational function* (the quotient of two polynomials) is continuous at each point where the denominator is nonzero.

Proof. Consider the rational function $f(x) = p(x)/q(x)$, where p and q are polynomials. Note that f is the product of p and the reciprocal of q. Suppose $q(x_0) \neq 0$. But the reciprocal of q is the composition of q followed by reciprocation, which was shown to be continuous in example 7 in §4.4. We have in fact proved much more:

Corollary B. *The limit of a quotient is the quotient of the limits, provided the limit of the denomiator is not zero,* in symbols,

$$\lim_{x \to x_0} \frac{f(x)}{g(x)} = \frac{\lim_{x \to x_0} f(x)}{\lim_{x \to x_0} g(x)}, \qquad (5.7)$$

provided $\lim_{x \to x_0} g(x) \neq 0$.

5.2 The Derivative

The two principal tools of calculus are the derivative and the integral. Here is the first.

Definition B. Suppose $f : X \subset \mathbf{R} \longrightarrow \mathbf{R}$ and suppose not only that is $x_0 \in X$ but that X contains an open neighborhood of x_0. Then the limit

---

[1] Repeat this mantra: *The limit of a sum is the sum of the limits. The limit of a product is the product of the limits.*

of the *difference quotient*

$$L = \lim_{x \to x_0} \frac{f(x) - f(x_0)}{x - x_0}, \tag{5.8}$$

if it exists, is called the *derivative of f* at x_0. The value L of this limit is written with any one of the symbols

$$Df\big|_{x_0}, \quad \frac{df}{dx}(x_0), \quad \frac{df}{dx}\bigg|_{x_0}, \quad f'(x_0), \tag{5.9}$$

and in application, when the independent variable is time t, it is common throughout science and engineering to use *Newton's notation*

$$\dot{y}(t_0) = \frac{dy}{dt}\bigg|_{t_0}. \tag{5.10}$$

Example 3. Constant functions $f(x) = c$ are everywhere differentiable with derivative 0.

Example 4. Power functions $f(x) = x^n$ with $n \in \mathbf{N}$ are everywhere differentiable with derivative given by the formula

$$Dx^n = nx^{n-1}. \tag{5.11}$$

To see this, note that

$$
\begin{aligned}
Df\big|_{x_0} &= \lim_{x \to x_0} \frac{f(x) - f(x_0)}{x - x_0} = \lim_{x \to x_0} \frac{x^n - x_0^n}{x - x_0} \\
&= \lim_{x \to x_0} \frac{(x - x_0)(x^{n-1} + x^{n-2}x_0 + x^{n-3}x_0^2 + \cdots + x_0^{n-1})}{x - x_0} \\
&= \lim_{x \to x_0} (x^{n-1} + x^{n-2}x_0 + x^{n-3}x_0^2 + \cdots + x_0^{n-1}) = nx_0^{n-1}. \tag{5.12}
\end{aligned}
$$

Theorem B. Suppose that both f and g are differentiable at $x = x_0$. Then so are cf, $f + g$, and fg, and their derivatives at x_0 are given by the rules

$$D(cf) = cDf, \tag{5.13}$$
$$D(f + g) = Df + Dg, \tag{5.14}$$
$$D(fg) = gDf + fDg. \tag{5.15}$$

Proof. The constant and addition rules follow directly from the limit rules (exercise 5.10). As for the product rule, we use standard trick #1:

$$D(fg)\big|_{x_0} = \lim_{x \to x_0} \frac{f(x)g(x) - f(x_0)g(x_0)}{x - x_0}$$

$$= \lim_{x \to x_0} \frac{f(x)g(x) - f(x_0)g(x) + f(x_0)g(x) - f(x_0)g(x_0)}{x - x_0}$$

$$= \lim_{x \to x_0} \frac{f(x) - f(x_0)}{x - x_0} g(x) + \lim_{x \to x_0} f(x_0) \frac{g(x) - g(x_0)}{x - x_0}$$

$$= \lim_{x \to x_0} \frac{f(x) - f(x_0)}{x - x_0} \lim_{x \to x_0} g(x) + f(x_0) \lim_{x \to x_0} \frac{g(x) - g(x_0)}{x - x_0}$$

$$= (\text{ detail }) = f'(x_0)g(x_0) + f(x_0)g'(x_0).$$

The missing detail is that *differentiability implies continuity.* That is, if g is differentiable at x_0, then g is continuous at x_0 (exercise 5.11). Hence $\lim_{x \to x_0} g(x) = g(x_0)$.

Corollary. Polynomials are everywhere differentiable with derivative formula

$$D(c_n x^n + c_{n-1} x^{n-1} + \cdots + c_1 x + c_0)$$
$$= c_n n x^{n-1} + c_{n-1}(n - 1)x^{n-2} + \cdots + c_1. \qquad (5.16)$$

Example 5. $D x^2 \tan x = 2x \tan x + x^2 D \tan x.$

Theorem C. (chain rule) Suppose f is differentiable at x_0 and g is differentiable at $f(x_0)$. Then $h = g \circ f$ is differentiable at x_0 with derivative at x_0 given by

$$D(g \circ f) = (Dg \circ f)Df, \qquad (5.17)$$

that is,

$$h'(x_0) = g'(f(x_0))f'(x_0). \qquad (5.18)$$

Defective proof. By standard trick #2,

$$Dg(f(x))\big|_{x_0} = \lim_{x \to x_0} \frac{g(f(x)) - g(f(x_0))}{x - x_0}$$

$$= \lim_{x \to x_0} \frac{g(f(x)) - g(f(x_0))}{f(x) - f(x_0)} \cdot \frac{f(x) - f(x_0)}{x - x_0} = g'(f(x_0))f'(x_0).$$

Critique and correct this proof as exercise 5.6.

Example 6. $D(1 + x^3)^7 = 7(1 + x^3)^6\, 3x^2$.

Example 7. As we shall soon see,

$$D \operatorname{Arctan} x = \frac{1}{1 + x^2}.$$

Hence

$$D \operatorname{Arctan} x^3 = \frac{1}{1 + (x^3)^2} Dx^3 = \frac{3x^2}{1 + x^6}.$$

Theorem D. (quotient rule) Suppose both f and g are differentiable at x_0 and that $g(x_0) \neq 0$. Then the quotient f/g is differentiable at x_0 with derivative at $x = x_0$ given by

$$D\frac{f}{g} = \frac{gf' - fg'}{g^2}. \tag{5.19}$$

Proof. The quotient map $h = f/g$ is the product of f with the composition of g followed by reciprocation $r(x) = 1/x$. Thus we need only prove r is differentiable everywhere but at $x = 0$ with formula

$$D\frac{1}{x} = -\frac{1}{x^2}. \tag{5.20}$$

But this is easy: For $x_0 \neq 0$,

$$D\frac{1}{x}\bigg|_{x_0} = \lim_{x \to x_0} \frac{1/x - 1/x_0}{x - x_0} = -\lim_{x \to x_0} \frac{1}{xx_0} = -\frac{1}{x_0^2}.$$

Example 8.

$$D\frac{x^3 - x + 4}{x^2 + 1} = \frac{(x^2 + 1)(3x^2 - 1) - (x^3 - x + 4)(2x)}{(x^2 + 1)^2}.$$

5.3 Mean Value Theorem

As we shall see in the next chapter, the following lemma is an extremely practical observation.

Lemma A. Suppose f is defined on an open set U containing c and differentiable at $x = c$. If f is maximum at $x = c$ on U, that is, $f(x) \leq f(c)$ for all $x \in U$, then

$$f'(c) = 0. \tag{5.21}$$

As an epigram, *the derivative vanishes at interior local extrema.*

Proof. Assume $f'(c) \neq 0$. Let F be the difference quotient

$$F(x) = \frac{f(x) - f(c)}{x - c}.$$

Since the limit

$$L = \lim_{x \to c} F(x)$$

exists, if its value L is nonzero, then by exercise 5.8, F must be of constant sign on some punctured neighborhood $0 < |x - c| < \delta$ contained in U. But since $f(x) - f(c) \leq 0$ on U, the difference quotient F must be nonnegative in U to the left of c and nonpositive to the right.

The following result, theorem C, is the third of the four pillars of calculus.

Theorem C. (mean value theorem) Suppose f is continuous on $[a, b]$ and differentiable on (a, b). Then for at least one intermediate point $a < c < b$,

$$\frac{f(b) - f(a)}{b - a} = f'(c). \tag{5.22}$$

Proof. Let

$$G(x) = f(x) - f(a) - \frac{f(b) - f(a)}{b - a}(x - a). \tag{5.23}$$

Note that $G(a) = G(b) = 0$. Because G is continuous, it must achieve its extreme values somewhere on $I = [a, b]$. If either extreme value is achieved at an interior point $x = c$, then by lemma A,

$$G'(c) = 0 = f'(c) - \frac{f(b) - f(a)}{b - a},$$

which is (5.22). On the other hand, if both extreme values occur at the endpoints, then $G(x) \equiv 0$, giving that $G(c) = 0$ for *every* interior point c.

Corollary A. *Functions with the same derivative differ by a constant.* That is, if $f'(x) = g'(x)$ on an open interval I, then for some constant C,

$$f(x) = g(x) + C \tag{5.24}$$

for all $x \in I$. As an epigram, *antiderivatives are determined within a constant.*

Proof. Set $h(x) = f(x) - g(x)$ and notice that $h'(x) \equiv 0$ on I. Fix $x_0 \in I$. Then for every $x \neq x_0$ in I, there is some intermediate c between x and x_0 for which

$$h'(c) = \frac{h(x) - h(x_0)}{x - x_0}.$$

But $h'(c) = 0$. Thus $h(x) \equiv h(x_0)$.

Corollary B. *Functions with positive derivatives are strictly increasing.* That is, if $f'(x) > 0$ on the open interval I, then on I,

$$x_1 < x_2 \quad \text{implies} \quad f(x_1) < f(x_2). \tag{5.25}$$

Proof. By the mean value theorem, $f(x_2) - f(x_1) = (x_2 - x_1)f'(c) > 0$.

Remark. By a slight modification of the last proof, we may see that *functions with nonnegative derivatives are nondecreasing, functions with negative derivatives are strictly decreasing,* and *functions with nonpositive derivatives are nonincreasing.*

Example 9. Since $f(x) = x^7 + 3x^5 + x - 2$ has the everywhere positive derivative $f'(x) = 6x^6 + 15x^4 + 1$, we see f is an everywhere increasing bijective map of $\mathbf{R}$ onto $\mathbf{R}$.

5.4 Derivatives of Inverse Functions

Lemma B. *Bijective continuous maps on compact sets have continuous inverses.*

Proof. Suppose $f : X \longrightarrow Y$ is a continuous bijective map from the compact set X onto the (hence) compact set Y. By Theorem C of §4.2, f maps closed, hence compact, subsets of X onto compact, hence closed, subsets of Y. That is to say, images of relatively open sets under f are relatively open.

Example 10. Consider any power function $f(x) = x^n$, $n \in \mathbf{N}$. There are two cases. For n even, since $f'(x) > 0$ on $[0, \infty)$, f is strictly increasing and a bijective map of $X = [0, \infty)$ onto itself. By lemma B, f^{-1} is continuous on each $[0, b^n]$. Thus the radical functions $\sqrt{x}, x^{1/4}, x^{1/6}, \ldots$ are all continuous bijective functions on the nonnegative reals.

Similarily, the odd radical functions $x^{1/3}, x^{1/5}, x^{1/7}, \ldots$ are all continuous bijective functions from $\mathbf{R}$ onto $\mathbf{R}$.

The inverse function theorem. Suppose f is continuous on $[a, b]$, differentiable on (a, b), with a nonzero derivative of one sign. Then the inverse function f^{-1} is differentiable on the interior of $[c, d] = f([a, b])$ with the value of the derivative at $y_0 = f(x_0)$ given by

$$Df^{-1}(y_0) = \frac{1}{f'(x_0)} = \frac{1}{f'(f^{-1}(y_0))}. \tag{5.26}$$

Proof. As we know from §4.2, f must map the closed and bounded interval $[a, b]$ onto a closed and bounded interval $[c, d]$. Because f has (say) a positive derivative on (a, b), it must be strictly increasing and so $f(a) = c$ and $f(b) = d$. Generically writing $y = f(x)$, we see

$$\frac{f^{-1}(y) - f^{-1}(y_0)}{y - y_0} = \frac{x - x_0}{f(x) - f(x_0)} = \frac{1}{F(x)} = \frac{1}{F(f^{-1}(y))},$$

where F is the difference quotient at x_0:

$$F(x) = \frac{f(x) - f(x_0)}{x - x_0}.$$

We may make F continuous at x_0 by setting $F(x_0) = f'(x_0)$. By lemma B, f^{-1} is continuous at y_0 and hence by exercise 5.12,

$$\lim_{y \to y_0} F(f^{-1}(y)) = f'(f^{-1}(y_0)) = f'(x_0).$$

Corollary A. The rational power function $f(x) = x^\alpha$, $\alpha \in \mathbf{Q}$, has derivative

$$Dx^\alpha = \alpha x^{\alpha-1} \tag{5.27}$$

at each point $x = x_0$ where $x_0^{\alpha-1}$ is defined.

Proof. We need only to establish the case $\alpha = 1/n$ for $n \in \mathbf{N}$; the complete result (5.27) will then follow via the chain rule (exercise 5.13). At $x_0 > 0$ for n even, or $x_0 \neq 0$ for n odd, by theorem D

$$Dx^{1/n}\Big|_{x=x_0} = \frac{1}{Dx^n\big|_{x=x_0^{1/n}}} = \frac{1}{nx^{n-1}\big|_{x=x_0^{1/n}}}$$

$$= \frac{1}{n}(x_0^{-1/n})^{n-1} = \frac{1}{n}x_0^{\frac{1}{n}-1}.$$

Example 11. The calculation $D\sqrt{x} = 1/2\sqrt{x}$ is valid for $x > 0$ while the following calculation is valid everywhere:

$$D\sqrt{x^2 + 1} = D(x^2 + 1)^{1/2} = (1/2)(x^2 + 1)^{-1/2}D(x^2 + 1) = x/\sqrt{x^2 + 1}.$$

5.5 Derivatives of Trigonometric Functions

Let us take a moment to recall the foundation of trigonometry. In a later chapter we will be able to carefully lay this foundation analytically, but here we must be content with an appeal to geometric intuition.

Consider the *unit circle* in $\mathbf{R}^2$, the locus of all points (x, y) satisfying $x^2 + y^2 = 1$. Beginning at $(1, 0)$, lay out a distance counterclockwise along the circumference of length $\theta > 0$. The x- and y-coordinate of the result is denoted by $\cos \theta$ and $\sin \theta$, respectively, as shown in figure 5.2.

For $\theta \leq 0$, lay out the distance $|\theta|$ in a clockwise direction. There are several immediately obvious geometric facts:

$$\cos^2\theta + \sin^2\theta = 1, \tag{5.28a}$$

$$\cos(-\theta) = \cos \theta, \qquad \sin(-\theta) = -\sin \theta, \tag{5.28b}$$

$$\cos(\theta + 2k\pi) = \cos \theta, \quad \sin(\theta + 2k\pi) = \sin \theta, \qquad k \in \mathbf{Z}, \tag{5.28c}$$

$$\cos(\theta + \pi) = -\cos \theta, \quad \sin(\theta + \pi) = -\sin \theta, \tag{5.28d}$$

$$\cos(\pi/2 - \theta) = \sin \theta, \quad \sin(\pi/2 - \theta) = \cos \theta, \tag{5.28e}$$

$$\lim_{\theta \to 0} \cos \theta = 1, \qquad \lim_{\theta \to 0} \sin \theta = 0. \tag{5.28f}$$

The following fundamental fact, however, is not at all obvious.

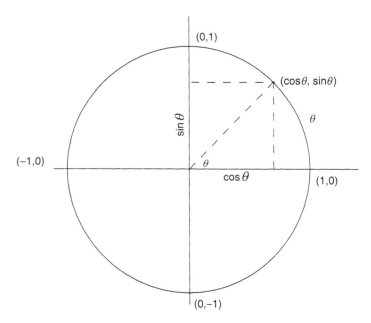

Figure 5.2 An angle θ that a ray from the origin makes with the x-axis is defined as the portion of the circumference of a unit circle cut out (measured counterclockwise). The cosine and sine of this angle are the x- and y-coordinates, respectively, of the intersection of this ray with the unit circle.

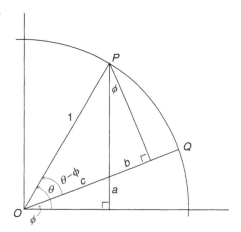

Figure 5.3 Two points P and Q are chosen in the first quadrant at angles θ and ϕ, respectively.

Lemma C. $\cos(\theta - \phi) = \cos\theta \cos\phi + \sin\theta \sin\phi.$ $\qquad$ (5.29)

Proof. Let us do the special case at the crux of the matter: $0 < \phi < \theta < \pi/2$. Examine figure 5.3.

By the trigonometric definitions and similarity, we have the relations

$$b = (\sin\theta - a)\sin\phi \quad \text{and} \quad \cos(\theta - \phi) = c + b, \qquad (5.30a)$$

$$\frac{a}{c} = \sin\phi \quad \text{and} \quad \frac{a}{\cos\theta} = \frac{\sin\phi}{\cos\phi}. \qquad (5.30b)$$

The lemma now follows (exercise 5.15).

Corollary.

$$\cos(\theta \pm \phi) = \cos\theta \cos\phi \mp \sin\theta \sin\phi, \qquad (5.31a)$$

$$\sin(\theta \pm \phi) = \sin\theta \cos\phi \pm \cos\theta \sin\phi, \qquad (5.31b)$$

$$\cos^2\theta = \frac{1 + \cos 2\theta}{2}, \quad \sin^2\theta = \frac{1 - \cos 2\theta}{2}. \qquad (5.31c)$$

Proof. Exercise 5.16.

And now, to the calculus of trigonometry.

Theorem D. Both sine and cosine are everywhere differentiable with differentiation formulas

$$D\sin x = \cos x \quad \text{and} \quad D\cos x = -\sin x. \qquad (5.32)$$

Proof. Since $\cos x = \sin(\pi/2 - x)$, the derivative formula for cosine follows from the differentiation formula for sine via the chain rule and (5.28e).

As for sine, by (5.31b),

$$
\begin{aligned}
D \sin x|_{x_0} &= \lim_{x \to x_0} \frac{f(x) - f(x_0)}{x - x_0} = \lim_{x \to x_0} \frac{\sin(x - x_0 + x_0) - \sin x_0}{x - x_0} \\
&= \lim_{x \to x_0} \frac{\sin(x - x_0) \cos x_0 + \cos(x - x_0) \sin x_0 - \sin x_0}{x - x_0} \\
&= \cos x_0 \lim_{x \to x_0} \frac{\sin(x - x_0)}{x - x_0} - \sin x_0 \lim_{x \to x_0} \frac{1 - \cos(x - x_0)}{x - x_0}.
\end{aligned}
$$

So to finish it is enough to show the famous limit

$$
\lim_{x \to x_0} \frac{\sin(x - x_0)}{x - x_0} = 1, \tag{5.33}
$$

from which follows (exercise 5.18)

$$
\lim_{x \to x_0} \frac{1 - \cos(x - x_0)}{x - x_0} = 0.
$$

We isolate the required and important result (5.33) as a separate lemma.

Lemma D. For small angles, $\sin \theta \approx \theta$. That is,

$$
\lim_{\theta \to 0} \frac{\sin \theta}{\theta} = 1. \tag{5.34}
$$

Proof. It is enough to show the result for small $\theta > 0$, since $\mathrm{sinc}\, x = \sin x / x$ is an *even* function; that is, $f(-x) = f(x)$. Examine figure 5.4.

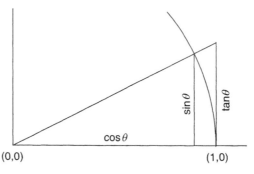

Figure 5.4 The angle θ cuts out three areas: an inner triangle, a sector, and an outer triangle.

Note that by comparing areas of the inner triangle, the sector, and the outer triangle, we have the geometrically inspired result

$$\frac{\cos\theta\sin\theta}{2} \le \frac{\theta}{2} \le \frac{\tan\theta}{2}.$$

Dividing through by $\sin\theta$ and clearing the factor of 2 yields

$$\cos\theta \le \frac{\theta}{\sin\theta} \le \frac{1}{\cos\theta}.$$

Taking limits as $\theta \to 0$, gives the equality (5.34) by the "squeeze lemma" of exercise 5.22.

Definition C. The most common ancillary trigonometric functions used nowadays[2] are, where defined,

$$\tan x = \frac{\sin x}{\cos x}, \quad \cot x = \frac{\cos x}{\sin x}, \tag{5.35a}$$

$$\sec x = \frac{1}{\cos x}, \quad \csc x = \frac{1}{\sin x}. \tag{5.35b}$$

Corollary. The ancillary differentiation formulas are, where defined,

$$D\tan x = \sec^2 x, \qquad D\cot x = -\csc^2 x, \tag{5.36a}$$

$$D\sec x = \sec x \tan x, \quad D\csc x = -\csc x \cot x. \tag{5.36b}$$

Proof. These formulas follow easily from the quotient rule (exercise 5.23).

Exercises

5.1   Prove that the notion of a limit of definition A in §5.1 is equivalent to the technical $\epsilon-\delta$ statement (5.2). (That is, prove that if the limit exists in one sense, then it does in the other.)

5.2   Prove that f is continuous on X if and only if

$$\lim_{x \to x_0} f(x) = f(x_0)$$

for every $x_0 \in X$.

5.3   Carefully prove (5.3).

[2]Several once-common ancillary functions, such as versin $\theta = 1 - \cos\theta$ and haversin $\theta = \sin^2(\theta/2)$, have vanished from use.

5.4   Carefully prove theorem A.

5.5   Prove that a rational function is continuous on all of **R** except
        possibly at a finite number of points.

5.6   What is wrong with the given proof of the chain rule (theorem C)?
        Write a correct proof.

5.7   Using a calculator, experimentally determine the limit

$$L = \lim_{x \to 0} \frac{\tan 2x \sec x}{\text{Arctan}(\sin 3x)}.$$

5.8   Prove that if $lim_{x \to x_0} f(x) = L \neq 0$, then $f(x)$ is of constant sign in
        some (punctured relative) neighborhood of x_0.
            Hint: Take $\epsilon = |L|$.

5.9   Carefully calculate the limit

$$L = \lim_{x \to x_0} \frac{x^3 - 2x^2 + x - 2}{x^3 - 3x^2 + 3x - 2}.$$

        Give a reason for each step.
            Alert: At $x_0 = 2$ (and only there) special care must be taken.

5.10   Prove (5.13) and (5.14).

5.11   Prove that if g is differentiable at x_0, then g is continuous at x_0.
            Hint: The difference quotient must be bounded near x_0.

5.12   Prove that if g is continuous at x_0 and if F is continuous at $y_0 = g(x_0)$
        and defined on an open neighborhood of y_0, then

$$\lim_{x \to x_0} F(g(x)) = F(g(x_0)).$$

5.13   Complete the proof of corollary A of §5.4.

5.14   Differentiate $f(x) = x^3/2 - 2x^4/7$. Where is your result valid?

5.15   Prove the formulas of (5.30), then deduce (5.29).

5.16   Deduce (5.31) from (5.29).

5.17   Prove the *Law of Cosines:* For a triangle of sides a, b, c, we have
        $c^2 = a^2 + b^2 - 2ab \cos \gamma$, where γ is the angle opposite c.

5.18   Deduce from (5.34) the limit

$$\lim_{\theta \to 0} \frac{1 - \cos \theta}{\theta^2} = \frac{1}{2}$$

and hence

$$\lim_{\theta \to 0} \frac{1 - \cos \theta}{\theta} = 0.$$

Hint: Multiply and divide by $1 + \cos \theta$.

5.19   Prove that the area A of a sector of a disc of radius r with central angle θ is

$$A = \frac{r^2 \theta}{2}.$$

5.20   Prove **Taylor's theorem:** If each of the higher derivatives f, f', f'', ..., $f^{(n+1)}$ exist at each point of an open interval I about $x = a$, then for each $b \in I$ there is a point c intermediate between b and a so that

$$f(b) = f(a) + f'(a)(b - a) + f''(a)\frac{(b - a)^2}{2!} + f'''(a)\frac{(b - a)^3}{3!}$$
$$+ \cdots + f^{(n)}(a)\frac{(b - a)^n}{n!} + f^{(n+1)}(c)\frac{(b - a)^{n+1}}{(n + 1)!}.$$

Outline: Aping the proof of the mean value theorem (MVT), let

$$G(x) = f(x) - f(b) + \sum_{k=1}^{n} f^{(k)}(x)\frac{(b - x)^k}{k!} + K\frac{(b - x)^{n+1}}{(n + 1)!},$$

where K is chosen so that $G(a) = 0$. Apply the MVT to G.

5.21   Find the first several terms of the Taylor expansion of $\cos x$ and $\sin x$.
        Answer:

$$\cos x = 1 - \frac{x^2}{2!} + \frac{x^4}{4!} - \frac{x^6}{6!} + \cdots$$
$$\sin x = x - \frac{x^3}{3!} + \frac{x^5}{5!} - \cdots.$$

5.22   Prove the "squeeze lemma": Suppose f and h have the same limit L at $x = x_0$ and that nearby x_0, $f(x) \le g(x) \le h(x)$. Then the limit of g exists at x_0 and has value L.

5.23   Verify the formulas of (5.36).

5.24   Prove that $f(x) = \tan x$ is a bijective map from $(-\pi/2, \pi/2)$ onto $\mathbf{R}$ and that its inverse function $f^{-1}(x) = \text{Arctan } x$ has derivative

$$D \text{ Arctan } x = \frac{1}{1 + x^2}.$$

Outline: Since $D \tan x = \sec^2 x > 0$, the formula (5.26) yields

$$D \text{ Arctan } x = \frac{1}{\sec^2(\text{Arctan } x)}.$$

But $\sec^2 x = 1 + \tan^2 x$.

5.25   Prove that $f(x) = \sin x$ is a bijective map from $[-\pi/2, \pi/2]$ onto $[-1, 1]$ and that its inverse function $f^{-1}(x) = \text{Arcsin } x$ has derivative

$$D \text{ Arcsin } x = \frac{1}{\sqrt{1 - x^2}}, \quad -1 < x < 1.$$

5.26   Prove that $f(x) = \cos x$ is a bijective map from $[0, \pi]$ onto $[-1, 1]$ and that its inverse function $f^{-1}(x) = \text{Arccos } x$ has derivative

$$D \text{ Arccos } x = -\frac{1}{\sqrt{1 - x^2}}, \quad -1 < x < 1.$$

5.27   Prove that $f(x) = |x|$ is not only everywhere continuous but differentiable everywhere but at $x_0 = 0$.

5.28   Construct a function that is everywhere continuous but fails to be differentiable at exactly the five points $x = 0, 1, 2, 3, 4$.
       Remark: Weierstrass exhibited an example of an everywhere continuous function that is nowhere differentiable. Can you guess how this is done? We shall see how to do this when armed with infinite series in chapter 9.

5.29   Prove *Darboux's Lemma:* If f is the derivative of a function g on (a, b), then f enjoys the intermediate value property on (a, b); that is, f maps any interval $I \subset (a, b)$ to an interval.
       Outline: Reduce to the case that if f changes sign, then it must possess a zero. Examine extrema of the continuous function g.

5.30   Produce an everywhere differentiable function whose derivative is continuous but not everywhere differentiable.

5.31   Can an everywhere differentiable function possess a discontinuous derivative?
        Suggestion: Consider $f(x) = x^2 \sin(1/x)$.

5.32   Prove the differentiation formula

$$D \sqrt{x} = \frac{1}{2\sqrt{x}}, \qquad x > 0,$$

        directly from definition B.
        Hint: Multiply and divide by the conjugate $\sqrt{x} + \sqrt{x_0}$.

5.33   Suppose that $f : X \subset \mathbf{R} \longrightarrow \mathbf{R}$ and that x_0 is either in X or a boundary point of $X \cap [x_0, \infty)$. We say f has the *right limit L* at $x = x_0$ and write

$$f(x_0^+) = \lim_{x \to x_0^+} f(x) = L$$

        if the limit of f on $X \cap [x_0, \infty)$ equals L. Left limits are defined in the obvious similar manner.
        Prove carefully that

$$\lim_{x \to 0^+} \operatorname{sgn} x = 1 \qquad \text{but} \qquad \lim_{x \to 0^-} \operatorname{sgn} x = -1,$$

        where the *signum* function $\operatorname{sgn} x = x/|x|$ if $x \neq 0$, 0 otherwise.

5.34   Prove that the limit of f exists at x_0 if and only if both left and right limits exist at x_0 and are equal.

5.35   Formulate an $\epsilon-\delta$ definition of right-hand limits.

5.36   Suppose f is defined on at least some half-open neighborhood $U = [x_0, x_0 + \delta)$ of x_0. We say f has the *right-hand derivative of value L* at $x = x_0$ if the right limit on U of the difference quotient $F(x) = (f(x) - f(x_0))/(x - x_0)$ exists and equals L. Left derivatives are defined in a similar way. Prove that $f(x) = (1 - x^2)^{3/2}$ is differentiable (in this now extended sense) with a continuous derivative on $[-1, 1]$.

5.37   **(Project)** Verify that all the limit theorems and derivative rules hold for left and right limits and derivatives.

5.38   Suppose $f : (a, \infty) \longrightarrow \mathbf{R}$. We say f has limit L at *positive infinity* and write

$$\lim_{x \to \infty} f(x) = L$$

if for every $\epsilon > 0$ there exists a $N > a$ so that

$$x > N \quad \text{implies} \quad |f(x) - L| < \epsilon.$$

The limit at $-\infty$ is defined in the obvious corresponding fashion. Prove that

$$\lim_{x \to \infty} \frac{1}{x} = \lim_{x \to -\infty} \frac{1}{x} = 0, \quad \lim_{x \to \infty} \frac{x}{1 + |x|} = 1, \quad \text{and} \quad \lim_{x \to -\infty} \frac{x}{1 + |x|} = -1.$$

5.39   Prove that all the limit rules hold when taking limits at ∞ or at $-\infty$.

5.40   Formulate an elegant topological definition of limit at ∞. Prove your definiton is equivalent to the $\epsilon-\delta$ definition of exercise 5.38.

5.41   Suppose $f : X \subset \mathbf{R} \longrightarrow \mathbf{R}$ and that x_0 is either in X or a boundary point of X. We write

$$\lim_{x \to x_0} f(x) = \infty$$

if for every $\epsilon > 0$ there exists a $\delta > 0$ so that for all $x \in X$, $0 < |x - x_0| < \delta$ implies $f(x) > 1/\epsilon$. Prove carefully that

$$\lim_{x \to 0} \frac{1}{|x|} = \infty.$$

5.42   Prove that the limit at ∞ of a rational function is determined solely by the relative degrees of its numerator and denominator, and if these degrees are equal, by the leading coefficients:

$$\lim_{x \to \infty} \frac{a_m x^m + \text{lower degree terms}}{b_n x^n + \text{lower degree terms}} = \begin{cases} \text{sgn}\,(a_m b_n)\,\infty & \text{if} \quad m > n \\ a_m/b_n & \text{if} \quad m = n \\ 0 & \text{if} \quad m < n. \end{cases}$$

Rules: You may use the results of any previous theorems or exercises.

5.43   In chapter 8 we will construct the *logarithm,* a bijective function $\ln : (0, \infty) \longrightarrow \mathbf{R}$ with $\ln 1 = 0$ and derivative

$$D \ln x = \frac{1}{x}.$$

Prove that the function exp inverse to logarithm has the simplest of all differentiation formulas: $D \exp x = \exp x$. Prove that $\exp(\alpha + \beta) = (\exp \alpha) \cdot (\exp \beta)$ and that $\exp(n\alpha) = (\exp \alpha)^n$ for all $\alpha, \beta \in \mathbf{R}, n \in \mathbf{Z}$.
    Notation. We will henceforth abbreviate $\exp x = e^x$.

5.44   **(Project)** Trace the history of the limit concept. Did it preceed or trail the concept of the derivative. Can the *avocat* Pierre de Fermat truly be credited with the discovery (invention) of the derivative?

5.45   Practice the differentiation rules of theorems B–D and the differentiation formulas of this chapter by calculating the following derivatives:

a. $D \dfrac{x^3 - 5x^2 + 7}{x^2 - x + 1}$

b. $D \dfrac{\ln \operatorname{Arcsin} x}{\csc x}$

c. $D \sqrt{x} \sin \sqrt{x}$

d. $D \sqrt{1 + \sqrt{x}}$

e. $D x^x$

f. $D \cos e^{\operatorname{Arctan} x}$

g. $D e^{\tan^3 x^2}$

h. $D 2^{1/x}$

i. $D \sqrt{\operatorname{Arcsin} x}.$

j. $D \cos^3 e^{x^2}.$

Specify for which x your calculations are valid.

5.46   Practice antidifferentiation by calculating

a. $D^{-1} \dfrac{\ln x}{x}$

b. $D^{-1} \dfrac{x}{1 + x^4}$

c. $D^{-1} \dfrac{x^3}{1 + x^4}$

d. $D^{-1} x^3 \sin x^4$

e. $D^{-1} \dfrac{x}{\sqrt{1 - x^2}}$

f. $D^{-1} x \cos x^2 \sin^5 x^2.$

5.47   Spend some time in introspection. To which camp do you belong? Do you believe that we *discover* mathematical truths (a Platonist), or do you believe these ideas are *invented* (a nominalist)? Compose a careful argument to buttress your position.

5.48   Why is positive angular direction counterclockwise?
          Hint: Think astronomy.

6

Applications of the Derivative

We will now survey the many uses of the derivative in geometry, sensitivity analysis, approximation, linearization, optimization, rates of change, mechanics, orbital motion, and economics. Herein are many valuable tools and techniques to be mastered.

6.1 Tangents

Suppose $f : [a, b] \longrightarrow \mathbf{R}$ and $a < x_0 < b$. What does it mean that "the graph of $y = f(x)$ has a tangent line $y = mx + b$ at $p = (x_0, f(x_0))$?" One may propose definitions such as "the tangent line is the line that meets the graph at p and only at p." But the example $f(x) = |x|$ at $x_0 = 0$ busts this first proposal—many lines meet the the graph of f only once at $(0, 0)$.

We are attempting to capture and generalize the familiar notion of a tangent to a circle. That suggests finding the best fitting circle to the graph of $y = f(x)$ at p, then using the tangent to this *osculating* circle as the tangent to the graph.

Or, appealing to modern experience, use the zoom feature of a graphing calculator to repeatedly zoom in toward the point p. (Try this with say $f(x) = x^2$ at $x_0 = 1$.) Notice how the apparent curvature of the graph of f flattens until the graph becomes a straight line—this is our tangent line.

But what has happened in this zooming process? We have been examining the portion of the curve $y = f(x)$ with x restricted to ever-smaller neighborhoods $(x_0 - \delta, x_0 + \delta)$, where we can no longer distinguish the curve from the *secant lines* of figure 6.1, whose slopes are given by difference quotients

$$F(x) = \frac{f(x) - f(x_0)}{x - x_0}. \tag{6.1}$$

This allows us to leap to the following.

Definition A. Suppose f is differentiable at $x = x_0$. Then the *line tangent to the graph of* $y = f(x)$ *at* $p = (x_0, f(x_0))$ is $y = mx + b$, where the slope m

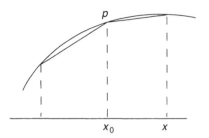

Figure 6.1 The secant lines from p. As we zoom in, the graph becomes indistinguishable from its tangent line.

is the derivative of f at x_0 and where the y-intercept b is chosen so that the line passes through p:

$$y = f(x_0) + f'(x_0)(x - x_0). \tag{6.2}$$

Example 1. Consider the graph of the function $f(x) = \operatorname{Arctan} x$ at the point $p = (1, \pi/4)$. Since $f'(1) = 1/(1 + 1^2) = 1/2$, the line tangent to $y = \operatorname{Arctan} x$ at p has equation

$$\frac{y - \pi/4}{x - 1} = \frac{1}{2},$$

that is,

$$y = \frac{x}{2} + \frac{\pi - 2}{4}.$$

Example 2. What is the point p of closest approach on the curve $y = x^2$ to the point $q = (2, 1)$? Intuitively (see exercise 6.18), the closest approach occurs where the line from q to p is normal to the curve, that is, when the point $p = (x, x^2)$ satisfies (exercise 6.3)

$$\frac{x^2 - 1}{x - 2} = -\frac{1}{2x}.$$

That is, the closest approach occurs at a point p with x-coordinate satisfying

$$2x^3 - x - 2 = 0, \tag{6.3}$$

giving that the closest approach occurs when $x \approx 1.165$ (exercise 6.1).

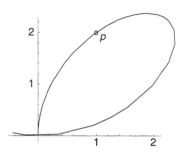

Figure 6.2 A portion of the *folium of Descartes*. The tangent to this curve at (1, 2) can be obtained via implicit differentiation, since locally the equation is the graph of a function.

Example 3. The famous *folium of Descartes* is the locus of all points satisfying an equation of type

$$x^3 + y^3 = 3axy. \tag{6.4}$$

See figure 6.2. Take the case $a = 3/2$. Note that the point $p = (1, 2)$ now lies on the curve. What is the line tangent to this curve at p?

This curve of figure 6.2 is *not* the graph of a function, since many vertical lines meet the curve more than once. However, the graph indicates that if we remain nearby p, the local portion of the folium through p appears to be given by a function; that is, the geometry predicts that we may solve locally (in theory)[1] for $y = y(x)$ as a function of x.

But then, placing $y = y(x)$ into (6.4) we obtain the identity $x^3 + y(x)^3 = 9xy(x)/2$, which we may differentiate to obtain via the chain and product rules an equation for y':

$$3x^2 + 3y^2y' = 9(xy' + y)/2. \tag{6.5}$$

Specializing at $p = (1, 2)$ gives $3 + 12y' = 9(y'+2)/2$, which we may solve for the slope of our sought-for tangent line to obtain $m = y'(1) = \frac{4}{5}$. Therefore our tangent line is

$$\frac{y - 2}{x - 1} = \frac{4}{5}.$$

This is the method of *implicit differentiation* where we may calculate $y'(x)$ without ever explicitly solving for y in terms of x.

---

[1] This geometric intuition is verified by one of the workhorses of advanced calculus— the *implicit function theorem*. See [MacCluer, 2005].

6.2 Newton's Method

Recall the bisection root-finding method from §4.4. It was slow but sure. The derivative provides far faster root-finding methods, all fraught with possible disaster. The most famous of these methods is credited to Newton.[2]

The Idea: At any x_0 near the zero crossing, by following the tangent line at $(x_0, f(x_0))$ to its intersection with the x-axis we may obtain an even closer approximation to the zero—see figure 6.3.

Newton's method. *Assumptions: f differentiable on $[a, b]$ and $f(a) \cdot f(b) < 0$.*
  Goal: To estimate a root of $f(x) = 0$ in $[a, b]$.

Pseudocode:

```
x = x₀, (some initial guess)
loop
   x = x − f(x)/f'(x)
until
   |f(x)| < required accuracy.
```

In short, repeatedly replace x with

$$x_{\text{next}} = x - \frac{f(x)}{f'(x)}. \tag{6.6}$$

Example 4. Let us again look at $f(x) = x^3 + x - 1$. As we saw in §4.4, because $f(0) \cdot f(1) < 0$, there must be a root of $f(x) = 0$ somewhere within $[0, 1]$. Start with some initial guess, say $x_0 = 1/2$. Note that $f'(x) = 3x^2 + 1$.

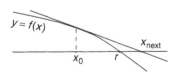

Figure 6.3 In Newton's algorithm, the intersection of the tangent line with the x-axis is the next approximation to the zero r of f.

---

[2]It is often observed that the most visible scientist of an era gets credit for all the work done in that era. See [Gleick].

First iteration. $x_0 = 0.5$.

$$x_1 = x_0 - f(x_0)/f'(x_0) = x_0 - (x_0^3 + x_0 - 1)/(3x_0^2 + 1)$$
$$= 1/2 - ((1/8)^3 + (1/2) - 1)/(3(1/2)^2 + 1) = 0.714285714.$$

Second iteration. $x_1 = 0.714285714$.

$$x_2 = x_1 - f(x_1)/f'(x_1) = 0.683179724.$$

Third iteration. $x_2 = 0.683179724$.

$$x_3 = x_2 - f(x_2)/f'(x_2) = 0.682328423,$$

which is good to six places. Moreover, $f(x_3) = 0.000001484$.

Example 5. Let us invent an algorithm for taking square roots by hand. Our goal is to find the positive root of

$$f(x) = x^2 - a = 0, \qquad a > 0. \tag{6.7}$$

By Newton, the recursion is

$$x_{\text{next}} = x - \frac{x^2 - a}{2x} = \frac{x + a/x}{2}, \tag{6.8}$$

that is, *average the divisor with the dividend.* So for instance when $a = 2$, take $x_0 = 3/2$ as the initial guess to the actual value $\sqrt{2} = 1.414213562 \cdots$ and apply our square-root algorithm:

First iteration. $x_0 = 3/2$.

$$x_1 = \frac{3/2 + 4/3}{2} = \frac{17}{12} = 1.41666 \cdots,$$

(already good to two places).

Second iteration. $x_1 = 17/12$.

$$x_2 = \frac{17/12 + 24/17}{2} = 1.414215687,$$

good to five places!

Rule of thumb. When convergent, *each iteration of Newton's method doubles the places of accuracy.* See exercise 6.5. That is the good news. However, there is bad news.

Alert. Newton's method can fail to converge.

Example 6. Let us search for the obvious zero $x = 0$ of $f(x) = x^{1/3}$ via Newton's method. At each iteration,

$$x_{\text{next}} = x - \frac{x^{1/3}}{(1/3)x^{-2/3}} = x - 3x = -2x,$$

and so the result of each iteration has twice the error of the preceding.

Remark A. The failure of Newton's algorithm to converge is not that unexpected in example 6, since f is not differentiable at its zero. But what conditions will guarantee convergence? Is there a basin of attraction about the zero that, once entered, will necessarily lead to the correct zero? This is indeed often the case. See exercise 6.4.

6.3 Linear Approximation and Sensitivity

Let us harken back to the original (ca. 1665) intuition surrounding the derivative—the notion of *infinitesimals*. If x is changed by an "infinitesimal" amount dx, then $y = f(x)$ is changed by some infinitesimal amount dy. The ratio dy/dx measures the *sensitivity* of f to small changes in x. Connecting their intuition to our limit viewpoint leads to the definition

$$dy = f'(x)dx. \tag{6.9}$$

So for example, the sensitivity of the circumference of a circle to changes in its radius r is constantly 2π. In contrast, the sensitivity of its area A to changes in r is $2\pi r$, thus area is more sensitive to change for larger r.

Example 7. What is the *relative* sensitivity of the volume V of a cube to an uncertancy of 2% in its side length x? Since $V = x^3$, we have

$$\frac{dV}{V} = \frac{3x^2 \, dx}{x^3} = 3\frac{dx}{x},$$

that is, a 2% change in side length can yield a 6% change in volume.

Note that this 6% relative sensitivity is only an approximation—a change from side length x_0 to x produces the actual relative change of

$$\frac{\Delta V}{V} = \frac{x^3 - x_0^3}{x_0^3} = \frac{x^2 + xx_0 + x_0^2}{x_0^2} \cdot \frac{x - x_0}{x_0}.$$

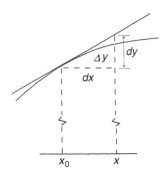

Figure 6.4 A close-up view of the tangent above x_0.

Remark B. We are employing an important back-of-the-envelope computational simplification widely employed in science and engineering: *For many local computations, we may replace a complicated function by its linear approximation. That is, its graph can be replaced by its tangent line.* In symbols,

$$f(x) \approx f(x_0) + f'(x_0)(x - x_0), \qquad (6.10)$$

or more compactly (see figure 6.4)

$$\Delta f \approx df. \qquad (6.11)$$

Example 8. Let us estimate $\sqrt{10}$ by linearizing $f(x) = \sqrt{x}$ at $x = 9$. Since $f'(x) = 1/2\sqrt{x}$,

$$\sqrt{10} \approx \sqrt{9} + \frac{1}{2\sqrt{9}} \cdot (10 - 9) = 3 + \frac{1}{6} = 3.1666\cdots,$$

which is good to two places.

6.4 Optimization

Recall from §4.2 that a continuous function f on a closed and bounded interval $[a, b]$ achieves its extreme values. And from §5.3, if f takes on an extreme value at an interior point $x = c$, then this point is *critical*; that is, either f fails to be differentiable at $x = c$ or $f'(c) = 0$.

Example 9. Assume that the *demand* for a commodity is $D(p) = 3 - p$, where $q = D(p)$ is the number of units that will be bought at price p. What price yields the most *revenue* $R = pq$?

Solution: Note that $R = pq = p(3 - p) = 3p - p^2$ and $0 \leq p \leq 3$. Thus revenue R will maximize either at the endpoints (where $R = 0$) or at an interior point where $D'(p) = 3 - 2p$, and hence the price yielding the largest revenue is $p = 1.5$.

Example 10. We wish to design an oven of volume $V = 10$ with a square base. The oven loses heat from its top per unit area at twice the rate it loses heat from the sides and at four times the rate it loses heat from the bottom. What is the most energy-efficient design shape?

Solution: Suppose the oven has dimensions x by x by y high. In the standard jargon of optimization, our problem is formalized as

Constraint: $x^2 y = 10$
Objective: To minimize $L = x^2 + x^2/4 + 4xy/2$.

We use the constraint to eliminate y from the objective function to obtain

$$L = \frac{5x^2}{4} + \frac{20}{x}, \quad 0 < x < \infty. \tag{6.12}$$

We then set the derivative of L equal to zero:

$$L' = \frac{10x}{4} - \frac{20}{x^2} = 0, \tag{6.13}$$

yielding but the one critical point $x = 2$.

This is an open endpoint problem, where we no longer can merely check values of the objective function at critical and endpoints to discover the extrema. As far as we know at this point, the objective L need not ever achieve its minimum on $0 < x < \infty$.

In such open endpoint problems we employ a standard trick called the *first derivative test:* Observe the sign patterns of the derivative L' in (6.13) of the objective function L. The sign of L' must remain constant between critical points, hence (since $L'(1) < 0$) constantly negative on $(0, 2)$ and (since $L'(3) > 0$) positive on $(2, \infty)$. Thus L is decreasing on $(0, 2)$, increasing thereafter—the minimum of L does indeed occur at $x = 2$. Therefore the most energy-efficient oven has dimensions 2 by 2 by 2.5.

6.5 Rate of Change

The consequences of this interpretation of the derivative are—without exaggeration—cosmic. Suppose $y(t)$ is a measurement of some quantity at time t, say of position, velocity, population, revenue, or temperature. Then *the derivative*

$$\dot{y}(t) = \frac{dy}{dt} \tag{6.14}$$

is the instantaneous rate of change of the quantity y at time t.

Example 11. It is observed that bodies very close to the surface of a planet fall with more or less constant acceleration g, that is, $\dot{v} = -g$, where v is vertical velocity measured upwards from the surface. But then, since *functions with the same derivative differ by a constant,* antidifferentiation gives $v = -gt + v_0$, where v_0 is the *initial* velocity (at $t = 0$). But $\dot{s} = v$, where s is the height above the surface. Again antidifferentiating, we have the *(near-planet) falling-body formula:*

$$s = -(1/2)gt^2 + v_0 t + s_0, \tag{6.15}$$

where s_0 is the initial height $s(0)$.

Example 12. A particle undergoing horizontal *harmonic motion*

$$x = \sin(\omega t - \phi) \tag{6.16a}$$

has velocity

$$v = \dot{x} = \omega \cos(\omega t - \phi) \tag{6.16b}$$

and acceleration

$$a = \dot{v} = -\omega^2 \sin(\omega t - \phi). \tag{6.16c}$$

Example 13. A population of size $p = p(t)$ at time t that is experiencing a growth rate proportional to its present population, that is, $\dot{p} = kp$, is, as we shall see, doubling exponentially: $p = p_0 2^{t/a}$, where a is the first doubling time and p_0 the initial population.

6.6 Related Rates

Example 14. (The beach-party problem) Suppose you and your friends are having a beach party one mile down the beach from the nearest approach of the beach to a lighthouse that is one mile offshore. See figure 6.5. The light is revolving counterclockwise (as they all do) when

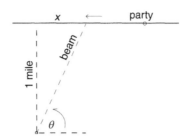

Figure 6.5 A lighthouse beam rotating at 2 rpm sweeps by a party one mile down the beach. The lighthouse is on a reef one mile offshore.

viewed from above at 2 rpm. At what speed does the light sweep past your party?

Solution: Let $x = x(t)$ be where the light first touches the beach at time t, measured as in figure 6.5. Then $x = \cot\theta$ and so $\dot{x} = -\dot{\theta}\csc^2\theta$. But because the light is revolving at 2 rpm, $\dot{\theta} = 4\pi$, and so at your location (where $\theta = \pi/4$), the beam of light sweeps by at $\dot{x} = -4\pi\csc^2\pi/4 = -4\pi\sqrt{2}$ (mi/min).

6.7 Ordinary Differential Equations

The power of science derives from its ability to predict the future and reconstruct the past. These predictions are by-in-large accomplished with *differential equations.*

Example 15. (The submarine problem) An attack submarine is cruising at forty knots at a depth of a thousand feet when suddenly the reactor scrams.[3] After one minute, way has dropped to thirty knots. How long does the crew have to make repairs before forward motion falls below steerageway of two knots?

Solution: Since the work necessary to move through a fluid is exchanged for the kinetic energy of the disturbed fluid particles, *the (force of) water resistance R against such a craft is more or less proportional to the square of its velocity.* That is,

$$R = -kv^2. \tag{6.17}$$

---

[3] A reactor shutdown is called a *scram.* This acronym came into use at the first reactor at the University of Chicago, for the job title of the "safety control rod axe man," who stood ready to cut the rope that held back the damping rods.

(The negative sign reflects that the resistance is in opposition to the velocity v.) We now apply *Newton's rule:*

$$\textit{sum of the inertial forces} = \textit{sum of the external forces.} \tag{6.18}$$

The inertial forces are one in number, namely $F = ma = m\dot{v}$, where m is the mass of the submarine. The external forces are also one in number, viz. $R = -kv^2$. Equating, we obtain our model:

$$m\frac{dv}{dt} = -kv^2 \tag{6.19}$$

subject to the data $v(0) = 40$, $v(1) = 30$. We ask for what t is $v(t) = 2$?
   Lump the unknown constants into one constant K to obtain

$$-\frac{\dot{v}}{v^2} = K. \tag{6.20}$$

Thus since functions with the same derivative differ by a constant, antidifferentiation yields

$$\frac{1}{v} = Kt + C. \tag{6.21}$$

Using our initial datum $v(0) = 40$, we see that $C = 1/40$, that is,

$$\frac{1}{v} = Kt + \frac{1}{40}. \tag{6.22}$$

Using our second datum $v(1) = 30$, we see that $K = 1/30 - 1/40 = 1/120$. All future velocities are thus predicted by

$$\frac{1}{v} = \frac{t}{120} + \frac{1}{40}. \tag{6.23}$$

Placing steerageway $v = 2$ into (6.23) yields $t = 57$, and so fifty-six minutes remain before steerageway is lost.

Remark C. Observe how the submarine problem above tracks exactly what some wags call the "scientific method":

   Step 1. Model the phenomenon as a differential equation.
   Step 2. Solve the differential equation.
   Step 3. Impose the given data.
   Step 4. Interpret the results.

6.8 Kepler's Laws

Johannes Kepler (ca. 1611) distilled from astronomical data amassed by Tycho Brahe three simple statements about the motions of the planets within our solar system.

Kepler I. Planetary motion is planar and in fact elliptical. See figures 6.6 and 6.7. In symbols,

$$x = r \cos \theta, \qquad\qquad y = r \sin \theta, \qquad\qquad (6.24)$$

with polar equation (exercise 6.19)

$$r = \frac{\gamma}{1 + e\cos \theta}. \qquad\qquad (6.25)$$

Kepler II. Planets sweep out equal area in equal time.

Sector area is proportional to sector angle (figure 6.8), that is, $dA/\pi r^2 = d\theta/2\pi$, and so $dA = r^2 d\theta/2$. But this means Kepler II implies $dA/dt = c$, giving that

$$r^2 \dot{\theta} = 2c \qquad\qquad (6.26)$$

and hence

$$2\dot{r}\dot{\theta} + r\ddot{\theta} = 0. \qquad\qquad (6.27)$$

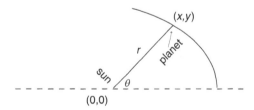

Figure 6.6 A planet can be located either by its Cartesian coordinates (x, y) or by its polar coordinates r, θ.

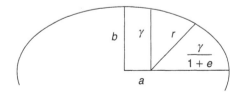

Figure 6.7 The orbit is an ellipse with semimajor axis a, semiminor axis b, and eccentricity e.

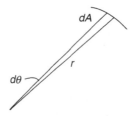

Figure 6.8 Sector area is proportional to sector angle.

Kepler III. The cube of the length a of the semimajor orbital axis of each planet is proportional to the square of its period p of rotation, that is,

$$a^3 = kp^2, \tag{6.28}$$

where the constant of proportionality k is identical for all planets.

Consequences of Kepler's laws. Differentiating both x and y in (6.24) twice with respect to time t and applying (6.27) yields (exercise 6.20)

$$\ddot{x} = (\ddot{r} - r\dot{\theta}^2) \cos \theta \tag{6.29a}$$
$$\ddot{y} = (\ddot{r} - r\dot{\theta}^2) \sin \theta. \tag{6.29b}$$

Thus *acceleration is centrally directed toward the sun.*
   Differentiating (6.25) gives (exercise 6.21)

$$\dot{r} = \frac{\gamma^2}{(1 + e \cos \theta)^2} \cdot \frac{\dot{\theta}}{\gamma} \cdot e \sin \theta$$
$$= r^2 \cdot \frac{e}{\gamma} \cdot \sin \theta \cdot \frac{2c}{r^2} = \frac{2ce}{\gamma} \sin \theta. \tag{6.30}$$

Therefore (exercise 6.22)

$$\ddot{r} = \frac{2ce\dot{\theta} \cos \theta}{\gamma} = \frac{4c^2 e}{\gamma} \cdot \frac{\cos \theta}{r^2}. \tag{6.31}$$

But by (6.26),

$$r\dot{\theta}^2 = r \left(\frac{2c}{r^2} \right)^2 = \frac{4c^2}{r^3}. \tag{6.32}$$

Thus (exercise 6.23)

$$\ddot{r} - r\dot{\theta}^2 = \frac{4c^2 e}{\gamma} \cdot \frac{\cos\theta}{r^2} - \frac{4c^2}{r^3} = \frac{4c^2}{r^2}\left[\frac{e\cos\theta}{\gamma} - \frac{1}{r}\right]$$

$$= \frac{4c^2}{r^2}\left[\frac{e\cos\theta}{\gamma} - \left(\frac{1 + e\cos\theta}{\gamma}\right)\right] = -\frac{4c^2}{\gamma} \cdot \frac{1}{r^2};$$ (6.33)

that is, from (6.29) we have the *inverse square law*

$$(\ddot{x}, \ddot{y}) = -\frac{4c^2}{\gamma} \cdot \frac{1}{r^2}(\cos\theta, \sin\theta).$$ (6.34)

Since planets sweep out equal area in equal time,

$$pc = \pi ab.$$ (6.35)

But then (exercise 6.24)

$$\frac{4c^2}{\gamma} = \frac{4}{\gamma}\left(\frac{\pi ab}{p}\right)^2 = \frac{4\pi^2 a^2 b^2}{\gamma p^2} = \frac{4\pi^2}{\gamma}\frac{a^2 b^2}{a^3}k = \frac{4\pi^2 k}{\gamma}\frac{b^2}{a}$$ (6.36)

$$= \text{(exercise 6.25)} = \frac{4\pi^2 b^2 k}{a^2(1 - e^2)} = \text{(exercise 6.25)} = 4\pi^2 k.$$ (6.37)

Hence from (6.34) comes *universal gravitation*

$$(\ddot{x}, \ddot{y}) = -\frac{4\pi^2 k}{r^2}(\cos\theta, \sin\theta)$$ (6.38)

for each planet of our solar system.

Critique. What hidden assumptions weaken these results on planetary motion? For one, we have assumed that the planets do not perturb one another and in turn that the sun is unperturbed by its planets. Does gravitational attraction act instantly through any distance? How much adjustment must be made for relativistic effects and the warping of space by the mass of the sun? (The precession of the orbit of Mercury argues for relativistic corrections.)

6.9 Universal Gravitation

In view of (6.38), we postulate the following.

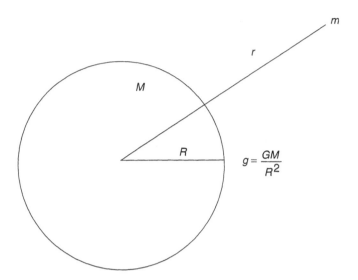

Figure 6.9 A radially symmetric planet of mass M acts like a point mass, inducing an inverse-square acceleration on a second mass m. The acceleration at the surface of the planet is traditionally denoted by g.

Observed Fact.[4] Given two masses m and M in empty space, each will induce an attractive acceleration upon the other along the line segment connecting them with a magnitude proportional to the other's mass and inversely proportional to the square of their distance apart—see figure 6.9.

In symbols, the mass m experiences the acceleration

$$a = -\frac{GM}{r^2}, \tag{6.39a}$$

while the mass M experiences the acceleration

$$A = -\frac{Gm}{r^2}, \tag{6.39b}$$

where r is the distance between the masses, and where G is the *universal constant of gravitation*. Note that the *forces* balance:

$$ma = MA. \tag{6.40}$$

---

[4]An *observed fact* is a hypothesis that is consistent with a large body of experiments.

Note that we have tacitly assumed that each mass acts as if it were a point mass. (This is verifiable for masses with radially symmetric densities via a triple integral computation.)

Maximum altitude. Place the origin of a coordinate system at the center of the large body M so that by universal gravitation, (6.39a) becomes

$$\ddot{r} = -\frac{GM}{r^2}. \tag{6.41}$$

Multiplying through by $2\dot{r}$ ("Newton's trick"), we obtain a relation where both sides are perfect derivatives:

$$2\dot{r}\ddot{r} = -2GM\frac{\dot{r}}{r^2},$$

giving that

$$\dot{r}^2 = \frac{2GM}{r} + C. \tag{6.42}$$

Let

$$g = \frac{GM}{R^2} \tag{6.43}$$

be the acceleration experienced at the surface of the large body M of radius R. Launch m radially from this surface with an initial velocity v_0. Then in this application, (6.42) specializes to (exercise 6.28)

$$\dot{r}^2 = \frac{2GM}{r} + v_0^2 - \frac{2GM}{R} = \frac{2GM}{r} + v_0^2 - 2gR. \tag{6.44}$$

Maximum range r_{max} is reached when $\dot{r} = 0$, that is, when (exercise 6.29)

$$r_{max} = \frac{R}{1 - v_0^2/2gR}. \tag{6.45}$$

Example 16. Fire a battleship gun with muzzle velocity $v_0 = 3025$ ft/s straight up. Since the radius of the Earth is $R = 4,000$ miles, and since $g = 32$ ft/s^2,

$$r_{max} = \frac{4000 \cdot 5280}{1 - (3025)^2/(2 \cdot 32 \cdot 4000 \cdot 5280)} = 21{,}263{,}953 \text{ feet}, \tag{6.46}$$

that is, an *altitude* above the surface of $r_{max} - R = 143{,}953$ ft $\approx$ 27 miles.

Escape Velocity. To entirely escape the gravitational well of the mass M, the initial velocity v_0 must be large enough so that $\dot{r} > 0$ for all time t and r, that is, from (6.44) (exercise 6.30)

$$v_0 = v_{esc} = \sqrt{2gR}. \qquad (6.47)$$

Example 17. What is the escape velocity from the Earth?

Solution: $v_{esc} = \sqrt{2 \cdot 32 \cdot 4000/5280} = \sqrt{48.48}$ mi/s $= 25{,}067$ mph.

Black Holes. Can a mass M be so great, or collapse to such a small radius that the escape velocity exceeds the speed c of light? From (6.47) we have that the mass M must have collapsed past the *Schwarzschild radius* (exercise 6.38)

$$R_{Sch} = 2g_0 \left(\frac{R_0}{c} \right)^2, \qquad (6.48)$$

where g_0, R_0 are the original values before the collapse.

Example 18. Since the speed of light is $c = 186{,}000$ mi/s, the Schwarzschild radius for the Earth is

$$R_{Sch} = 2 \cdot 32 \cdot \left(\frac{4000}{186{,}000} \right)^2 = 0.03 \text{ ft} = 0.36 \text{ inches.}$$

In contrast, the Schwarzschild radius of the Sun is 1.85 miles (exercise 6.39).

6.10 Concavity

The second time-derivative $\ddot{s}$ of displacement s is acceleration, the rate of change of the velocity. But the second derivative with respect to a space variable signifies geometric *concavity*. If the second derivative is positive, then the first derivative must be increasing, hence the graph must be curving upward. Let us make this precise.

Definition C. The graph of a differentiable function f is locally *concave upward* at $x = x_0$ if for some open interval I containing x_0, each point of the graph of $y = f(x)$, $x \in I$, lies on or above the tangent line at $(x_0, f(x_0))$. See figure 6.10. Likewise, the graph of $y = f(x)$ is *concave downward* at $x = x_0$ if it is locally below the tangent line at $x = x_0$.

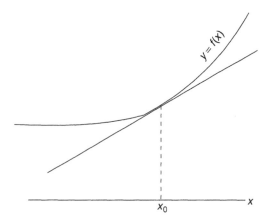

Figure 6.10 A curve is concave up at a point x_0 when the graph locally lies above the line tangent at x_0.

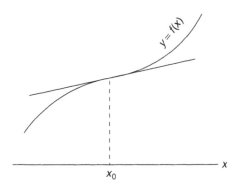

Figure 6.11 A point of inflection is where concavity changes, in this example from concave down to concave up.

Corollary. Suppose that f is differentiable on an open neighborhood of x_0 and that $f''(x_0)$ exists. Then

    a. If f is concave upward at $x = x_0$, then $f''(x_0) \geq 0$.

    b. If f is concave downward at $x = x_0$, then $f''(x_0) \leq 0$.

    c. If $f''(x_0) > 0$, then f is concave upward at $x = x_0$.

    d. If $f''(x_0) < 0$, then f is concave downward at $x = x_0$.

Proof. Exercise 6.44.

    If there has been a change in concavity at $x = x_0$ from concave up to concave down (or *vice versa*), then x_0 is a *point of inflection*, as in the example $y = x^3$ at $x_0 = 0$. See figure 6.11. When f'' is continuous

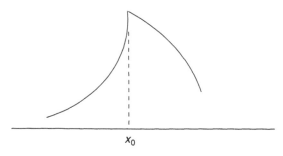

x_0

Figure 6.12 A point of inflection can also occur at
points of nondifferentiability.

near x_0, a change in concavity implies the vanishing of the second
derivative $f''(x_0)$.

However, the vanishing of $f''(x_0)$ can occur in many other situations.
The curve may in fact be concave up (or down) at x_0, yet the graph is
approximated by its tangent to the second order (as does the tangent
$y = 0$ to $y = x^4$ at $x_0 = 0$). There are other, more complicated situations
where the second derivative may vanish, yet no conclusions about
concavity can be drawn.

A change in concavity can also occur where the derivative fails to
exist, as at the critical point of figure 6.12.

Example 19. Let us construct a qualitative sketch of the graph of

$$f(x) = \frac{x}{x^2 - 1}. \tag{6.49}$$

Note that the graph of $y = f(x)$ has horizontal asymptote $y = 0$ as
$x \to \pm\infty$ and vertical asymptotes at $x = \pm 1$. Note also that

$$f'(x) = -\frac{x^2 + 1}{(x^2 - 1)^2}$$

and hence f is everywhere decreasing. Finally note that

$$f''(x) = \frac{2x(x^2 + 3)}{(x^2 - 1)^3}.$$

Since functions are of constant sign between their zeros or points
of discontinuity, f'' is of constant sign on each of the four intervals
$(-\infty, -1)$, $(-1, 0)$, $(0, 1)$, $(1, \infty)$. Testing at one point in each of these
four intervals, we see that f'' has sign pattern $-, +, -, +$, respectively.
Thus f is concave down, up, down, up, respectively, giving that $x = \pm 1$,
and 0 are points of inflection. A sketch of the graph of f is shown in

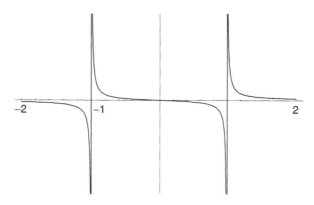

Figure 6.13 A sketch of the graph of (6.49).

figure 6.13. Half of this work was unnecessary, since f is *odd*, that is, $f(-x) = -f(x)$, and thus its graph is antisymmetric about the y-axis.

6.11 Differentials

What is the sensitivity of a function of many variables $f(x_1, x_2, \ldots, x_n)$ to a change in only one of its variables? It is the derivative of f with respect to this one variable while holding all other variables constant, that is, a *partial derivative*.

Example 20. The partial derivatives of $f(x, y, z) = xy^2z^3$ with respect to x, y, z, respectively, are

$$\frac{\partial f}{\partial x} = y^2z^3, \quad \frac{\partial f}{\partial y} = 2xyz^3, \quad \frac{\partial f}{\partial z} = 3xy^2z^2. \tag{6.50}$$

Question. What is the combined sensitivity of a function of many variables to small changes in all of its variables simultaneously? The surprising answer is that the effects of the individual changes merely superimpose.

Technical result. Suppose all of the partial derivatives $\partial f / \partial x_i$ are continuous on an open neighborhood of $x^0 = (x_1^0, x_2^0, \ldots, x_n^0) \in \mathbf{R}^n$. Then on some open ball about x^0,

$$f(x^0 + dx) - f(x^0) = df(x^0) + E, \tag{6.51a}$$

where the *total differential*

$$df(x^0) = \frac{\partial f(x^0)}{\partial x_1} dx_1 + \frac{\partial f(x^0)}{\partial x_2} dx_2 + \cdots + \frac{\partial f(x^0)}{\partial x_n} dx_n, \tag{6.51b}$$

and where the error E satisfies

$$\lim_{|dx| \to 0} \frac{E}{|dx|} = 0. \tag{6.51c}$$

Put into words, *except for "second-order" error, local change is given by the total differential.*

Proof. This result follows from the multivariable version of the mean value theorem. See [MacCluer, 2005].

Example 21. What is the relative sensitivity of the volume V of a rectangular box of dimensions 1 by 2 by 3 to changes of 0.4%, 0.5%, and 0.6% in its respective dimensions?

Solution. Since volume $V = xyz$,

$$\frac{dV}{V} = \frac{yz\,dx + xz\,dy + xy\,dz}{xyz} = yz\frac{dx}{x} + xz\frac{dy}{y} + xy\frac{dz}{z}$$

$$= 6\frac{dx}{x} + 3\frac{dy}{y} + 2\frac{dz}{z} = 6 \cdot 0.004 + 3 \cdot 0.005 + 2 \cdot 0.006$$

$$= 0.051 = 5.1\%.$$

Remark D. Since partial derivatives inherit the rules of ordinary derivatives, there is nothing new to learn in order to compute total derivatives. For example, by the product and chain rules, $dx \cos y^2 z = \cos y^2 z\,dx - (x \sin y^2 z)\,dy^2 z = \cos y^2 z\,dx - (x \sin y^2 z)(2yz\,dy + y^2\,dz)$.

Exercises

6.1   Using Newton's method, approximate the root r of $2x^3 - x - 2 = 0$ in $[1, 2]$ to three decimal places.
      Hint: Iterate until the third place is stable.
      Answer: $r = 1.16537304 \cdots$.

6.2   Without a calculator, estimate $\sqrt{10}$ to three decimal places.

6.3   Prove that any line normal (perpendicular) to the line $y = mx + b$, $m \neq 0$, has slope $n = -1/m$.

6.4   Assume f is twice-continuously differentiable at and near one of its zeros $x = r$. Assume also that $f'(r) \neq 0$. Prove that on all sufficiently small open neighborhoods U about $x = r$, any initial choice x_0 within U will converge under Newton's iteration to the zero r.

Outline: We may without loss of generality assume $r = 0$ and $f'(0) = 1$. Then for x near 0, by Taylor's formula (exercise 5.20),

$$|x_{\text{next}} - r| = \left| x - \frac{f(x)}{f'(x)} \right| = \left| x - \frac{x + f''(c)x^2/2}{1 + f''(d)x} \right|$$

$$= \left| \frac{x + f''(d)x^2 - x - f''(c)x^2/2}{1 + f''(d)x} \right| = \left| \frac{B}{1 + Cx} \right| \cdot |x|^2.$$

6.5 Deduce from exercise 6.4 the rule of thumb that each iteration of a converging Newton's algorithm doubles the places of accuracy.

6.6 (An old chestnut) Lay a rope around the equator of the Earth. Its length will be approximately 25,000 miles. Next raise the rope ten feet at all points. How much longer a rope is required? You will be astonished.

Hint: Use differentials.

6.7 Suppose we are to design a speed control for an automobile with speed dependence $m\dot{v} + kv = u$, where v is the velocity referenced to the set speed (say 65 mph), and $u = a\theta$, where θ is the accelerator pedal angle referenced to its nominal position at the set speed. (The kv term is the linearization of air resistance about the set speed.) We close the loop by designing a mechanical linkage that proportionally depresses the pedal when there is a sensed drop in velocity: $\theta = -cv$. Is this closed-loop system stable? That is, will the solutions $v = v(t)$ that have been perturbed away from 0 return to 0?

6.8 Astronomers estimate the distance to nearby stars by dividing their diameter (determined via a *main sequence* correlation of size with luminosity and color) by their subtended angle (in radians). What justifies this computation?

6.9 Implicitly compute the slope of the tangent to the unit circle $x^2 + y^2 = 1$ at each point (x, y), $y \neq 0$. Check your answer by solving explicitly for y and differentiating.

6.10 Compute implicitly from (6.5) the value of the second derivative $y''(1)$.

Answer: $y''(1) = -0.864$.

6.11   It is observed that *light will take the path of least time* (Fermat's principle). Suppose the speed of light in the upper half plane $y > 0$ is c_1, while the speed in the lower half plane $y < 0$ is c_2 as in the following figure.

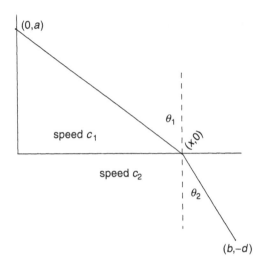

Calculate the total transit time T for light leaving the point $(0, a)$ to reach $(b, -d)$. Minimize T to obtain *Snell's Law*:

$$\frac{\sin \theta_1}{c_1} = \frac{\sin \theta_2}{c_2}.$$

6.12   Prove the *second derivative test:* Suppose that f is differentiable on an open neighborhood U of $x = c$, that $f'(c) = 0$, and that $f''(c)$ exists. Then $f''(c) < 0$ implies $x = c$ is a local maximum and $f''(c) > 0$ implies $x = c$ is a local minimum.

    Outline: First apply exercise 5.8 to the difference quotient $F(x) = (f'(x) - f'(c))/(x - c)$, then apply the first derivative test of §6.4.

6.13   The stiffness of a rectangular beam is proportional to the product of its width and the square of its depth. Find the stiffest beam that can be cut from a circular log.

6.14   A rectangle in the first quadrant with sides parallel to the coordinate axes has one vertex at the origin $(0, 0)$, the opposite vertex at (x, e^{-x}). For which x (if any) is the area of this rectangle maximum and minimum?

6.15   A policeman parked a hundred feet from a highway observes with
        radar that an oncoming motorist three hundred feet distant is closing
        at 59 mph. What is the speed of the oncoming car?
            Answer: 62.6 mph.

6.16   A directly approaching airplane is flying level at 40,000 feet. At one
        moment, when the elevation θ from you to the plane is 30°, you
        measure that θ is increasing at 0.25 degrees per second. Calculate the
        speed of the airplane.
            Answer: 476 mph.

6.17   A soda pop bottle is thrown upward at 40 feet per second from the
        edge of a roof of a two hundred foot tall building and falls to the
        pavement below. Calculate the maximum height reached and the
        time before impact.
            Answer: 225 ft., 5 sec.

6.18   Suppose that f is everywhere differentiable. Prove that the closest
        approach of the graph of f to an external point is a perpendicular
        distance.

6.19   An *ellipse* is the locus of all points P in the plane such that the sum of
        the distances from P to two fixed points Q and R is constant. Prove
        that when $Q = (-f, 0)$ and $R = (f, 0)$, the cartesian equation of this
        ellipse is of the form

        $$\frac{x^2}{a^2} + \frac{y^2}{b^2} = 1,$$

        where $b^2 + f^2 = a^2$. When the coordinate system is translated
        horizontally so that one focus lies at the origin, prove that the polar
        equation for the ellipse is now given by (6.25), where $a^2(1 - e^2) = b^2$
        and $\gamma = a(1 - e^2)$.

6.20   Verify (6.29).

6.21   Verify (6.30).

6.22   Verify (6.31).

6.23   Verify the string of calculations leading to (6.33).

6.24   Verify (6.36).

6.25   Verify (6.37) You may assume the elliptic geometric facts
        $a^2(1 - e^2) = b^2$ and $\gamma = a(1 - e^2)$.

6.26*   Conversely, deduce each of Kepler's laws from universal gravitation

$$\ddot{\mathbf{r}} = -\frac{GM}{r^2}\mathbf{u},$$

where $\mathbf{r} = (x, y, z)$, $r = \sqrt{x^2 + y^2 + z^2}$ and $\mathbf{u} = \mathbf{r}/r$.
Outline: Differentiate $\mathbf{p} = \mathbf{r} \times \dot{\mathbf{r}}$ to see that motion is planar. Show
that any planar central acceleration field will satisfy Kepler II by
revisiting your computations leading to (6.29). Make the
substitution $z = 1/r$ to find that $\dot{\theta} = 2cz$ and that $dr/dt = -2cdz/d\theta$
and hence $z'' + z = \pi^2 k/c^2$. Thus $r = \gamma/(1 + e\cos\theta)$. Finally, using
$cp = 4\pi ab$ and the geometric relations for an ellipse, deduce that
$a^3 = kp^2$.

6.27   Verify (6.42).

6.28   Verify (6.44).

6.29   Verify (6.45).

6.30   Verify (6.47).

6.31   **(Project)** Design experiments to determine the following
        Earth-relative masses and radii:

	M/M_e	R/R_e
Earth	1	1
Moon	0.0123	0.272
Mars	0.107	0.538
Jupiter	315	11.3
Sun	329,400	109

6.32   What is the gravitational constant g_m at the surface of the Moon?
        Standard trick: Using the table of Earth-relative mass and radius of
        exercise 6.31,

$$g_m = \frac{GM_m}{R_m^2} = \left(\frac{GM_e}{R_e^2}\right)\left(\frac{R_e}{R_m}\right)^2\frac{M_m}{M_e} = g_e\left(\frac{R_e}{R_m}\right)^2\frac{M_m}{M_e}.$$

Answer: 5.32 ft./sec^2.

6.33   What is the escape velocity from the Sun?

$$v_{esc} = \sqrt{2g_s R_s} = \sqrt{\frac{2GM_s}{R_s}}.$$

Answer: 1,378,005 mph.

6.34   The Moon is 250,000 miles distant. Where between the Earth and the Moon do their gravitational fields cancel?
Answer: 225,042 miles from the center of the Earth.

6.35   Can you hit Cleveland from the Moon using a battleship gun with muzzle velocity 3025 ft./sec.?

6.36   Is the Moon a satellite or a sister planet?
Hint: Where does their center of mass lie?

6.37   Assuming you can jump three-feet high here on Earth, how high could you jump on the Moon?
Answer: 18 feet.
Outline: If liftoff velocity is v_0, then by equating initial kinetic energy $T = mv_0^2/2$ at liftoff to eventual potential energy $V = mgh$ at maximum height h, we see[5] $h = v_0^2/2g$. But then

$$\frac{h_m}{h_e} = \frac{g_e}{g_m} \approx 6.$$

6.38   Verify (6.48).

6.39   Find the Schwarzchild radius of the Sun.
Answer: 1.85 miles.

6.40   **(Project)** Reconstruct the history of the discovery of the inverse square law. Is it true that Newton only realized a decade later that the inverse square law followed from Kepler's laws? It is said that he originally came to the result by experimenting with many rules like $\ln r, \sqrt{r}, 1/r, r^2, \ldots$ until he found that $1/r^2$ matched observations. Is this correct?

6.41   Using differentials, estimate the percentage decrease from sea level in the acceleration g at the summit of Mt. Everest (altitude 29,000 ft.).
Answer: 0.274%

[5]This can also be obtained more laboriously via (6.15)

6.42   (A standard method of obtaining inequalities) Prove that if
       $f'(x) \leq g'(x)$ on $[a, b]$, then $f(x) - f(a) \leq g(x) - g(a)$ on $[a, b]$.

6.43   Deduce from repeated applications of exercise 6.42 the inequality

$$1 - \frac{x^2}{2} \leq \cos x.$$

6.44   Prove the corollary to definition C in §6.10.
       Outline: To prove (a), note that for $x > x_0$, by the MVT, $f'(c) =$
       $(f(x) - f(x_0))/(x - x_0) \geq f'(x_0)$ for some $x_0 < c < x$. Thus $0 \leq$
       $\lim_{x \to x_0^+} (f'(c) - f'(x_0))/(x - x_0) = f''(x_0)$.
          To prove (c), suppose $f''(x_0) > 0$. Then by exercise 5.8, the
       difference quotient $F(x) = (f'(x) - f'(x_0))/(x - x_0)$ must be positive
       on some open interval I about x_0. Hence $x_1 < x_0 < x_2$ implies
       $f'(x_1) < f'(x_0) < f'(x_2)$. It now follows from exercise 6.42 that the
       graph locally lies above the tangent line at $x = x_0$ on I.

6.45   Sketch locally the graph of a function f with $f(x_0) = 1$, $f'(x_0) = -1$,
       and $f''(x_0) = 1$.

6.46   Decode the following econspeak: "Marginal costs over the coming
       fiscal year are anticipated to grow at a decreasing rate."

6.47   Show that a portion of the graph of $y = f(x)$ of constant *curvature*

$$\kappa = \frac{y''}{(1 + y'^2)^{3/2}}$$

       is a portion of a circle of radius $a = 1/|\kappa|$.

6.48   A steel beam is experiencing *shear stress* when forces are acting
       parallel to a cross section of the beam. Argue that the existence of
       shear forces is signaled by the change in the rate of change of the
       second derivative of the beam's shape.

6.49   By hand (without a graphing calculator) roughly sketch the graph of

$$y = \frac{x^2}{(x - 3)^2},$$

       showing all asymptotes, critical points, local extrema, concavities,
       and points of inflection.

6.50   Solve the differential equation $\dot{p} = kp$ of example 13 modeling
        unrestrained population growth. Employ exercise 5.43.

6.51   One day it began snowing at a slow and steady rate. At noon, a
        snowplow began plowing, clearing two miles the first hour, one mile
        the second. What time did it begin snowing?
            Answer: 11:23 a.m.
            Hint: Derive the volume rate of clearing in two different ways and
            equate.

6.52   Verify that each point on the folium of Descartes (6.4) can be given
        by the rational parameterization

$$x = \frac{3at}{1+t^3}, \qquad y = \frac{3at^2}{1+t^3}.$$

6.53   Prove that the kinetic energy $T = mv^2/2$ of a satellite of mass m
        in a circular orbit of radius r about a planet of mass M is $T = GMm/2r$.
            Hint: The orbital velocity can be deduced from Kepler III. Obtain
            the value for k in (6.28) by equating (6.38) and (6.39).

6.54   Solve the following initial value problems by separating variables:
            a. $\dot{y} = -2y$, $y(0) = 1$
            b. $\dot{y} = ty$, $y(0) = -1$
            c. $\dot{y} = 1 + y^2$, $y(0) = 0$
            d. $\ddot{y} = \dot{y}$, $y(0) = 1, \dot{y}(0) = -1$

6.55   A can of soda pop is taken from a 20°F refrigerator and placed on a
        countertop in a room at temperature 70°F and forgotten. One minute
        later the pop has warmed one degree. What is its temperature after
        sitting for one hour on the countertop?
            Answer: 55°F.
            Hint: By *Newton's law of cooling*, the rate of change of temperature
            T is proportional to the temperature difference driving the heat
            transfer, in this case, $\dot{T} = k(70 - T)$.

6.56   A murder victim is discovered floating in 50°F water with a body
        temperature of 72°F. Fifteen minutes later the body temperature has
        dropped two degrees. When did the murder occur?
            Answer: Approximately two hours before the discovery of the
            body.

6.57 A wire of length 1 is cut into two pieces, then bent to form a square and a circle. What division results in the least total area?

6.58 The *planar pendulum* of bob mass m and arm a

has kinetic energy $T = ma^2\dot{\theta}^2/2$ and potential energy $V = mga(1-\cos\theta)$. (Why?) Prove that the motion of this pendulum satisfies

$$\ddot{\theta} = -\frac{g}{a}\sin\theta.$$

Deduce that for *small* maximal deflections θ_0, the pendulum oscillates with period

$$p \approx 2\pi\sqrt{\frac{g}{a}}.$$

In fact, the actual period of a pendulum increases with its maximal deflection until, in the limit, $p = \infty$ when $\theta_0 = \pi/2$. See exercise 8.73.

6.59 Modern radios and televisions are tuned digitally using *phase-locked loops,* closed-loop electronic systems that lock onto programmed divisor frequencies of some master clock. The *channels* are the stable[6] constant solutions $\theta(t) = c$ (*equilibria*) of an ODE of the form $\dot{\theta} = -a\sin\theta$. Find all equilibria of this system. Which of these equilibria are channels?

---

[6]An equiliblium solution is *stable* when solutions that initiate nearby will tend asymptotically to the equilibrium.

7

The Riemann Integral

We now leave differential calculus and enter integral calculus. We construct the Riemann integral, prove the fundamental theorem, and investigate which functions are integrable.

7.1 Darboux Sums

Assume f is a bounded function defined on the closed and bounded interval $[a, b]$. Consider any *partition* $\mathcal{P}$ of $[a, b]$, that is, any finite set of points of $[a, b]$ that includes the two endpoints a, b. We think of this partition

$$\mathcal{P} : x_0 = a < x_1 < x_2 < \cdots < x_n = b \tag{7.1}$$

as dividing the interval into the subintervals $[x_{i-1}, x_i]$ of length $\Delta x_i = x_i - x_{i-1}$. Let the infimum and supremum of f on each subinterval be denoted by

$$m_i = \inf_{[x_{i-1}, x_i]} f(x) \quad \text{and} \quad M_i = \sup_{[x_{i-1}, x_i]} f(x). \tag{7.2}$$

The sums

$$L(\mathcal{P}) = m_1 \Delta x_1 + m_2 \Delta x_2 + \cdots + m_n \Delta x_n = \sum_{i=1}^{n} m_i \Delta x_i, \tag{7.3a}$$

$$U(\mathcal{P}) = M_1 \Delta x_1 + M_2 \Delta x_2 + \cdots + M_n \Delta x_n = \sum_{i=1}^{n} M_i \Delta x_i \tag{7.3b}$$

are called the *lower* and *upper Darboux sums of f*, respectively, *for the partition $\mathcal{P}$*.

Lemma A. *No lower sum can exceed any upper sum.* That is, if $\mathcal{P}_1$ and $\mathcal{P}_2$ are any two partitions of $[a, b]$, then

$$L(\mathcal{P}_1) \leq U(\mathcal{P}_2). \tag{7.4}$$

Proof. Let $\mathcal{P}_0$ be a common refinement of both partition $\mathcal{P}_1$ and partition $\mathcal{P}_2$. For example, form the refinement $\mathcal{P}_0 = \mathcal{P}_1 \cup \mathcal{P}_2$. But refinements can only increase the lower sum and decrease the upper (exercise 7.1). Thus

$$L(\mathcal{P}_1) \le L(\mathcal{P}_0) \le U(\mathcal{P}_0) \le U(\mathcal{P}_2). \qquad (7.5)$$

Thus *each lower sum is a lower bound for the set of all upper sums* and *each upper sum is an upper bound for the set of all lower sums.*

Corollary. *The supremum of all lower sums is at most the infimum of all upper sums.* In symbols,

$$\sup_{\mathcal{P}} L(\mathcal{P}) \le \inf_{\mathcal{P}} U(\mathcal{P}). \qquad (7.6)$$

Definition. If equality holds in (7.6), then the function f is said to be *Riemann integrable* over the interval $[a, b]$ and the common value of (7.6) is written with the symbol

$$\int_a^b f(x)\, dx = \sup_{\mathcal{P}} L(\mathcal{P}) = \inf_{\mathcal{P}} U(\mathcal{P}). \qquad (7.7)$$

More informally, if there is only one number trapped between all upper and lower sums, then the Riemann integral exists and equals this unique trapped number.

Example 1. (Lejeune-Dirichlet's example) Let f be defined by the rule

$$f(x) = \begin{cases} 1 & \text{if } x \in \mathbf{Q} \\ 0 & \text{otherwise.} \end{cases}$$

Consider partitions of $[0, 1]$. Because each subinterval of any partition contains both rationals and irrationals, all $m_i = 0$ and all $M_i = 1$, giving that $L(\mathcal{P}) = 0$ and $U(\mathcal{P}) = 1$ for all partitions $\mathcal{P}$. Thus f is not integrable over $[0, 1]$.

Example 2. Consider the function f given by $f(x) = x$ over the interval $[0, 1]$. For any partition $\mathcal{P} : 0 = x_0 < x_1 < x_2 < \cdots < x_n = 1$, because f is increasing,

$$L(\mathcal{P}) = \sum_{i=1}^{n} f(x_{i-1}) \Delta x_i$$

and

$$U(\mathcal{P}) = \sum_{i=1}^{n} f(x_i) \Delta x_i.$$

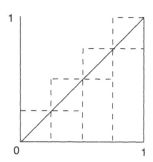

Figure 7.1 Because $f(x) = x$ is increasing, the lower sum is calculated using values of f at left end-point values, while the upper sum is calculated with values to the right.

See figure 7.1. Note that if the n subintervals of this partition are all of equal length $\Delta x = 1/n$, then

$$U(\mathcal{P}) - L(\mathcal{P}) \; = \; \frac{1}{n},$$

and so, by choosing n arbitrarily large we may force equality in (7.6) and thus f is integrable.

Moreover, the resulting upper sum is

$$U_n = \sum_{i=1}^{n} f(i/n)(1/n) = \frac{1}{n^2} \sum_{i=1}^{n} i = \frac{1}{n^2}[1 + 2 + 3 + \cdots + n]$$

$$= (\text{exercise } 7.2) = \frac{1}{n^2} \frac{n(n+1)}{2} = \frac{1}{2} \frac{n+1}{n}.$$

Hence

$$\inf_{n} U_n \; = \; \int_0^1 x \, dx \; = \; \frac{1}{2}. \tag{7.8}$$

7.2 The Fundamental Theorem of Calculus

The following stunning result intertwines differential and integral calculus. It is considered one of the milestones of European thought.

Theorem A. (The fundamental theorem of calculus) Suppose

   a. f is integrable on $[a, b]$,

   b. F is continuous on $[a, b]$,

   c. F is differentiable on (a, b), and

   d. $F'(x) \; = \; f(x)$ on (a, b).

Then

$$\int_a^b f(x)\, dx \ = \ F(b) - F(a). \tag{7.9}$$

Proof. Consider any partition $\mathcal{P} : a = x_0 < x_1 < \cdots < x_n = b$. By the mean value theorem there are $x_i^* \in (x_{i-1}, x_i)$ such that

$$F(b) - F(a) \ = \ \sum_{i=1}^n F(x_i) - F(x_{i-1})$$

$$= \ \sum_{i=1}^n F'(x_i^*)(x_i - x_{i-1}) \ = \ \sum_{i=1}^n f(x_i^*)\Delta x_i. \tag{7.10}$$

Hence the number $F(b) - F(a)$ is caught between all upper and lower sums. But since f is integrable, there is only one such number, namely

$$I \ = \ \int_a^b f(x)\, dx.$$

Example 3. The integrals of polynomials are trivial to obtain. For instance,

$$\int_{-1}^2 (3 - 2x + x^2 + x^3)\, dx \ = \ 3x - x^2 + \frac{x^3}{3} + \frac{x^4}{4} \bigg|_{-1}^2$$

$$= \ \left[3 \cdot 2 - 2^2 + \frac{2^3}{3} + \frac{2^4}{4} \right]$$

$$- \left[3(-1) - (-1)^2 + \frac{(-1)^3}{3} + \frac{(-1)^4}{4} \right].$$

Example 4.

$$\int_0^{\pi/2} \sin^4 x \cos x \, dx = \frac{\sin^5 x}{5} \bigg|_0^{\pi/2} = \frac{\sin^5 \pi/2}{5} - \frac{\sin^5 0}{5} = \frac{1}{5}.$$

7.3 Continuous Integrands

In examples 3 and 4 we have tacitly assumed the integrals exist, as verified by the following not-so-surprising result.

Theorem B. *Continuous functions are integrable.* That is, if f is continuous on $[a, b]$, then f is integrable over $[a, b]$.

Proof. Because the interval $[a, b]$ is compact, f is not only continuous, but is in fact *uniformly continuous* on $[a, b]$; that is, for every $\epsilon > 0$, there exists a $\delta > 0$ such that for all $x_0, x \in [a, b]$,

$$|x - x_0| < \delta \quad \text{implies} \quad |f(x) - f(x_0)| < \epsilon. \tag{7.11}$$

See lemma B below.[1]

Therefore, if we choose a partition $\mathcal{P}$ of $[a, b]$ into n equal subintervals each of length $\Delta x < \delta$, we see that the differences between the maximum and minimum values of f on each subinterval are less than ϵ, and so

$$U(\mathcal{P}) - L(\mathcal{P}) = \sum_{i=1}^{n} (M_i - m_i)\Delta x \leq \sum_{i=1}^{n} \epsilon \Delta x = \epsilon(b - a). \tag{7.12}$$

Thus the upper and lower sums are forced together and equality holds in (7.6); that is, f is integrable.

Lemma B. *A function continuous on a compact set is uniformly continuous.*

Proof. Let us prove the result for any function $f : X \to Z$ from the metric space X to the metric space Z. Suppose f is continuous on the compact subset K of X. Fix $\epsilon > 0$. Since f is continuous on K, for each $x \in K$ there is a $\delta(x) > 0$ so that for all $y \in K$,

$$d(x, y) < \delta(x) \quad \text{implies} \quad d(f(x), f(y)) < \epsilon/2. \tag{7.12}$$

The open balls with center x of radius $\delta(x)/2$ cover K, a cover that possesses a finite subcover of balls $B_1, B_2, \ldots, B_p$ centered at $x_1, x_2, \ldots, x_p$, respectively. Choose

$$\delta = \min_{1 \leq i \leq p} \delta(x_i)/2. \tag{7.13}$$

Suppose $x, y \in K$ and $d(x, y) < \delta$. Since the B_i cover K, the point $y \in B_1$ (say) and hence $d(y, x_1) < \delta(x_1)/2$. But then, since $d(x, y) < \delta \leq \delta(x_1)/2$, the point x must be at distance $d(x, x_1) < \delta(x_1)$. Hence by (7.12), $d(f(x), f(y)) \leq d(f(x), f(x_1)) + d(f(x_1), f(y)) < \epsilon/2 + \epsilon/2 = \epsilon$. That is, for every $\epsilon > 0$ there exists a $\delta > 0$ such that for all $x, y \in K$,

$$d(x, y) < \delta \quad \text{implies} \quad d(f(x), f(y)) < \epsilon. \tag{7.14}$$

---

[1]Notice the subtle difference between continuity and uniform continuity. Borrowing the standard quantification symbols from logic, the statement of continuity begins with $\forall \epsilon > 0 \; \forall x_0 \; \exists \delta > 0 \; \forall x \ldots$, while uniform continuity has the inner two quantifiers reversed: $\forall \epsilon > 0 \; \exists \delta > 0 \; \forall x_0 \; \forall x \ldots$. In words, the same δ works everywhere, for every x_0.

7.4 Properties of Integrals

Theorem C. If f and g are integrable over $[a, b]$, then so are $f + g$ and cf for any constant c. Moreover,

$$\int_a^b (f(x) + g(x))\, dx = \int_a^b f(x)\, dx + \int_a^b g(x)\, dx \qquad (7.15)$$

and

$$\int_a^b cf(x)\, dx = c \int_a^b f(x)\, dx. \qquad (7.16)$$

Proof. Exercise 7.4.

Theorem D. If f is integrable over $[a, b]$ then f is integrable over any subinterval $[a_1, b_1] \subset [a, b]$.

Proof. Let $\epsilon > 0$ be given. Let $\mathcal{P}$ be any partition of $[a, b]$ where the Darboux sums for f satisfy $U(\mathcal{P}) - L(\mathcal{P}) < \epsilon$. Refine this partition by adding the endpoints a_1, b_1. Let $\mathcal{P}_1$ be the partition of $[a_1, b_1]$ obtained by discarding points in the refinement not in $[a_1, b_1]$. Then the Darboux sums of f over $[a_1, b_1]$ for this partition satisfy $U(\mathcal{P}_1) - L(\mathcal{P}_1) < \epsilon$, since we have discarded nonnegative contributions.

Corollary. Suppose f is integrable over $[a, b]$. If $a < c < b$, then

$$\int_a^b f(x)\, dx = \int_a^c f(x)\, dx + \int_c^b f(x)\, dx. \qquad (7.17)$$

Proof. Exercise 7.5.

Theorem E. If f and g are integrable over $[a, b]$ and $f(x) \leq g(x)$ on $[a, b]$, then

$$\int_a^b f(x)\, dx \leq \int_a^b g(x)\, dx. \qquad (7.18)$$

Proof. Exercise 7.6.

Theorem F. If f is continuous on $[a, b]$, then

$$\int_a^b f(x)^2\, dx = 0 \quad \text{implies} \quad f(x) = 0 \quad \text{for all } x \in [a, b]. \qquad (7.19)$$

Proof. Exercise 7.7.

Notation. We extend the integral symbol by setting

$$\int_a^a f(x)\, dx = 0 \qquad (7.20)$$

and

$$\int_b^a f(x)\, dx = -\int_a^b f(x)\, dx. \qquad (7.21)$$

7.5 Variable Limits of Integration

Some refer to the following result as the "baby fundamental theorem of calculus."

Theorem G. If f is continuous on $[a, b]$, then

$$\frac{d}{dx}\int_a^x f(t)\, dt = f(x) \qquad (7.22)$$

at each $x \in [a, b]$.

Proof. Fix $x_0 \in [a, b]$ and set[2]

$$F(x) = \int_a^x f(t)\, dt. \qquad (7.23)$$

Then the difference quotient

$$\frac{F(x) - F(x_0)}{x - x_0} = \frac{1}{x - x_0}\left[\int_a^x f(x)\, dx - \int_a^{x_0} f(t)\, dt\right]$$

$$= \frac{1}{x - x_0}\int_{x_0}^x f(t)\, dt.$$

When $x_0 < x$, using a partition of $[x_0, x]$ of one subinterval,

$$f(x_m) \le \frac{1}{x - x_0}\int_{x_0}^x f(x)\, dx \le f(x_M),$$

where x_m and x_M are points between x_0 and x where f achieves its minimum and maximum values, respectively. (When $x_0 > x$ the inequalities reverse.) But in the limit (exercise 7.8),

$$\lim_{x \to x_0} f(x_m) = f(x_0) = \lim_{x \to x_0} f(x_M). \qquad (7.24)$$

---

[2]The logicians insist that we must use a "dummy variable" such as t for the bound variable of the integrand in (7.23) because the upper limit of integration is free to vary.

Corollary. Suppose f is continuous on an interval I containing all values of the differentiable functions u, v on $[a, b]$. Then

$$\frac{d}{dx} \int_{u(x)}^{v(x)} f(t)\, dt = f(v(x))v'(x) - f(u(x))u'(x) \qquad (7.25)$$

at each $x \in [a, b]$.

Proof. Fix $y_0 \in I$ and set

$$F(y) = \int_{y_0}^{y} f(t)\, dt.$$

Then by the chain rule, $F(v(x)) - F(u(x))$ has derivative $F'(v(x))v'(x) - F'(u(x))u'(x)$, which is our desired result.

Example 5.

$$\frac{d}{dx} \int_{\tan x}^{x^3} \cos t^2\, dt = 3x^2 \cos x^6 - \sec^2 x \cos \tan^2 x.$$

7.6 Integrability

Which functions are integrable? We have proved that continuous functions are integrable. Dirichlet's example 1 is everywhere discontinuous and not integrable. So where between continuous everywhere and continuous nowhere is the dividing line for integrability? A finite number of discontinuities is not serious.

Theorem H. *A bounded function with at most a finite number of discontinuities is integrable.*

Proof. Suppose the bounded function f is continuous on $[a, b]$ except at a finite number of points. We immediately reduce to the case of exactly one discontinuity by dividing up the interval $[a, b]$ into subintervals each containing only one discontinuity; for if f is integrable over each subinterval, it is integrable over the whole interval, by exercise 7.10, the converse to theorem D.

Let x^* be the point of discontinuity of f on $[a, b]$ but suppose $x^* \neq a, b$. Let $\epsilon > 0$ be given, and let $m = \inf_{[a,b]} f(x)$ and $M = \sup_{[a,b]} f(x)$. Choose a subinterval $I = (a_1, b_1) \subset [a, b]$ containing x^* satisfying

$$(M - m)(b_1 - a_1) < \epsilon/2.$$

The complement $[a, b] \setminus I$ consists of two disjoint closed intervals where f is continuous with a combined partition $\mathcal{P}$ such that in total,

$U(\mathcal{P}) - L(\mathcal{P}) < \epsilon/2$. But then the points of $\mathcal{P}$ form a partition of all of $[a, b]$ where

$$U(\mathcal{P}) - L(\mathcal{P}) < \epsilon.$$

A similar argument obtains when $x^* = a$ or b.

Remark. The crux of the proof of theorem H was to isolate the dicontinuities of f into subintervals so small that their contributions to the Darboux sums were negligible. Informally, if the set D of points of discontinuity can be contained in sets of arbitrarily small size, then f is integrable. That in fact is the complete answer: *A bounded Borel-measurable[3] function is Riemann integrable over a bounded closed interval $[a, b]$ if and only if the set of its discontinuities in $[a, b]$ has Lebesgue measure zero.*

Allow me to explain. Consider the collection $\mathcal{B}$ of *Borel* sets, the smallest collection of subsets of **R** that is closed under countable union and complementation that contains all open sets. It is a fundamental result of graduate analysis that there exists a natural *measure* on the Borel sets; that is, there is a real- or infinite-valued function μ with the three properties

I. $\mu(E) \geq 0$ for every Borel set E.

II. For any countable disjoint collection of Borel sets E_k,

$$\mu\left(\bigcup_{k=1}^{\infty} E_k\right) = \sum_{k=1}^{\infty} \mu(E_k).$$

III. $\mu([a, b]) = b - a$.

Thus there is a concept of "length" of Borel sets that agrees with the ordinary notion of length on intervals. If the "length" of the set of discontinuities of f is 0, then f is integrable and conversely. For the details of this famous result see [Bruckner et al.].

Exercises

7.1   Prove that when a partition is refined, (i.e., more points are added), the lower sum either increases or remains the same while the upper sum either decreases or remains the same.

7.2   Prove that $1 + 2 + 3 + \cdots + n = n(n+1)/2$.

---

[3]A function f is *Borel measurable* if the inverse image under f of every Borel set is again a Borel set—a very mild requirement.

7.3   **(Project)** Trace the history of the Riemann integral. Does Cavalieri
        deserve credit for the idea? Do not recent documents show
        Archimedes employed the concept? Who is credited with first
        uncovering the fundamental theorem? Why is this integral named
        after the much-later Georg Riemann?

7.4   Prove theorem C.
        Hint:

$$\inf_E f(x) + \inf_E g(x) \leq \inf_E (f(x) + g(x))$$

$$\leq \sup_E (f(x) + g(x)) \leq \sup_E f(x) + \sup_E g(x).$$

7.5   Prove the corollary to theorem D.

7.6   Prove theorem E.

7.7   Prove theorem F.
        Hint: If f is nonzero at a point it is nonzero on a neighborhood of
        that point.

7.8   Verify (7.24).

7.9   Show that $f(x) = x^2$ is integrable on $[0, 1]$ directly using Darboux
        sums. Find the value of the integral directly (without using the
        fundamental theorem).
        Hint: Show $1 + 4 + 9 + \cdots + n^2 = n(n+1)(2n+1)/6$.

7.10   Prove that if f is integrable over $[a, c]$ and over $[c, b]$, where $a < c < b$,
        then f is integrable over $[a, b]$.
        Outline: For $\epsilon > 0$, there are partitions $\mathcal{P}_1$ of $[a, c]$ with
        $U(\mathcal{P}_1) - L(\mathcal{P}_1) < \epsilon/2$. Likewise there are partitions $\mathcal{P}_2$ of $[c, b]$ where
        $U(\mathcal{P}_2) - L(\mathcal{P}_2) < \epsilon/2$. Thus for the partition $\mathcal{P} = \mathcal{P}_1 \cup \mathcal{P}_2$ of $[a, b]$, we
        have $U(\mathcal{P}) - L(\mathcal{P}) < \epsilon$.

7.11*   **(Riemann sums)** Let f be integrable over $[a, b]$. For any partition
        $\mathcal{P} : a = x_0 < x_1 < \cdots < x_n = b$ and any choices $x_i^* \in [x_{i-1}, x_i]$, a sum of
        the form

$$\sum_{i=1}^{n} f(x_i^*) \Delta x_i$$

        is called a *Riemann sum* of f. Prove that the Riemann sums of f
        eventually cluster about the integral of f; that is, prove that for every

$\epsilon > 0$ there exists a $\delta > 0$ such that

$$\max_{1 \le i \le n} \Delta x_i < \delta \quad \text{implies} \quad \left| \sum_{i=1}^{n} f(x_i^*) \Delta x_i - \int_a^b f(x)\, dx \right| < \epsilon.$$

7.12   Occasionally a sum can be recognized as a Riemann sum of a familiar integral. Calculate the limit

$$\lim_{n \to \infty} \sum_{k=1}^{n} \frac{3n^2 + 2k^2}{n^3}.$$

Answer: 11/3.

7.13**   Construct an example of a bounded function that is a derivative on $[a, b]$ yet is not integrable over $[a, b]$.

7.14*   Prove that if f is differentiable at x_0, then its *Lanczos derivative* at x_0 given by the limit

$$LDf(x_0) = \lim_{h \to 0} \frac{3}{2h^3} \int_{-h}^{h} t f(x_0 + t)\, dt$$

exists and equals $f'(x_0)$. In contrast, although $f(x) = |x|$ is not differentiable at $x_0 = 0$, prove that its Lanczos derivative does exist at $x_0 = 0$ and equals 0.

8

Applications of the Integral

We now learn many of the important uses of the Riemann integral in mathematics, science, and engineering. The results depend in most cases on physical/geometric intuition or observed facts, not on an axiomatic foundation. Thus these results have no proofs, only persuasive verifications.

8.1 Work

By definition, *work done while moving a distance d against a constant force f is* $W = f \cdot d$. But what if the force f is not constant?

Result A. Suppose a particle at location x experiences a force[1] $f(x)$, where f is continuous on $[a, b]$. Then the total work W done by this particle moving from a to b against the force f is

$$W = - \int_a^b f(x) \, dx. \tag{8.1}$$

Verification. Partition $[a, b]$ into small trips, $\mathcal{P} : a = x_0 < x_1 < \cdots < x_n = b$. The actual work W_i done moving the small distance $\Delta x_i = x_i - x_{i-1}$ from x_{i-1} to x_i is certainly trapped:

$$\inf_{[x_{i-1}, x_i]} f(x) \Delta x_i \leq -W_i \leq \sup_{[x_{i-1}, x_i]} f(x) \Delta x_i. \tag{8.2}$$

Thus the total actual work

$$W = \sum_{i=1}^n W_i$$

is caught between the upper and lower Darboux sums for any partition, giving (8.1).

---

[1] By convention, positive force is to the right, negative to the left. Thus moving to the right with a force to the right is negative work.

Remark A. Many of our verifications below will use this same idea—that the actual physical or geometric quantities are Riemann sums or at least caught between all upper and lower Darboux sums. However, this is not how most scientists and engineers verify these results. Their intuition harkens back to the early viewpoint of infinitesimals. For example, they would argue result A as follows:

During the movement from x through an infinitesimal distance dx, the force f changes little and so the contribution to the total work is $dW = -f(x)\, dx$. Summing over all these infinitesimal motions yields the total work

$$W = \int_a^b dW = -\int_a^b f(x)\, dx.$$

I recommend that you embrace this intuition. It is quick and effective and rarely leads you astray.

Example 1. A mass m is free to slide along a horizontal frictionless rail (see figure 8.1) restrained by an ideal spring of spring constant k. Hooke's law states that *the restoring force F of a spring is proportional to the displacement x from its natural length.* In symbols, $F = -kx$. Hence if the mass is displaced x from rest, the work done must be

$$W = -\int F\, dx = \int_0^x ky\, dy = kx^2/2. \tag{8.3}$$

While we are here, note that as another consequence of Hooke's law, since the mass m is experiencing only the restoring force of the spring (and inertial force must be balanced by external forces), its acceleration a obeys $ma = F = -kx$. Therefore the mass is undergoing *harmonic motion*; that is, displacement x satisfies the ODE

$$\ddot{x} + \omega^2 x = 0, \tag{8.4}$$

where $\omega^2 = k/m$.

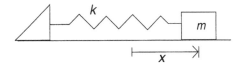

Figure 8.1 A spring-mass system.

8.2 Area

Result B. Suppose f is integrable and nonnegative on $[a, b]$. The area A of the region R bounded by the graph of $y = f(x)$, the x-axis, and the vertical lines $x = a$ and $x = b$ is

$$A = \int_a^b f(x)\, dx. \tag{8.5}$$

See figure 8.2.

Verification. Whatever one's notion of area, certainly the area A under the curve is caught between the upper and lower Darboux sums of any partition. See figure 8.2. But there is exactly one such real number, namely the value of the intergral (8.5).

Remark B. Those same scientists or engineers would instead argue that the total area is clearly the sum of all the infinitesimal areas $dA = f(x)\, dx$ with base dx and height $f(x)$. See figure 8.3. After all, they would point out, the elongated "S" integral symbol indicates it is a "smear" or "continuous" sum. See exercise 8.69.

Alert. The integral (8.5) gives the geometric (absolute) area provided $f \geq 0$. When the graph of f dips below the x-axis, the absolute area bounded by $y = f(x)$, $x = a$, $x = b$, and the x-axis is given by

$$A = \int_a^b |f(x)|\, dx. \tag{8.6}$$

To evaluate integrals of absolute values $|f|$, divide the integration into intervals where the sign of f is constant, then integrate $\operatorname{sgn}(f(x))f(x) = |f(x)|$.

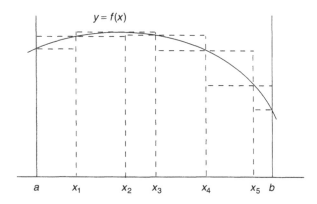

Figure 8.2 Whatever is the area A under curve $y = f(x)$, it must be trapped between the upper and lower sum.

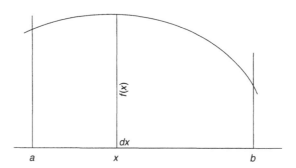

Figure 8.3 The area is the "continuous" or "smear" sum of the infinitesimal rectangles of base dx and altitude $f(x)$.

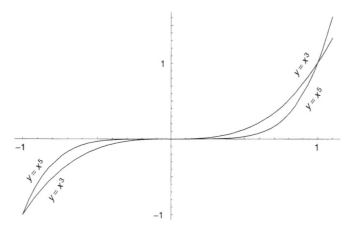

Figure 8.4 The graphs of $y = x^3$ and $y = x^5$ superimposed.

Likewise, the absolute area bounded by two curves $y = f(x)$ and $y = g(x)$ is given by

$$A = \int_a^b |f(x) - g(x)| \; dx. \tag{8.7}$$

Example 2. What is the area A of the region bounded by the curves $y = x^3$, $y = x^5$, $x = -1$, and $x = 2$?

Solution. A good graph is essential—see figure 8.4. The (bounded) region caught by the given curves has absolute area

$$A = \int_{-1}^{2} |x^3 - x^5| \; dx$$

$$= \int_{-1}^{0} x^5 - x^3 \; dx + \int_{0}^{1} x^3 - x^5 \; dx + \int_{1}^{2} x^5 - x^3 \; dx.$$

8.3 Average Value

Result C. The average value av(f) of an integrable function f on $[a, b]$ is given by

$$\mathrm{av}(f) \;=\; \frac{1}{b-a} \int_a^b f(x)\, dx. \qquad (8.8)$$

Verification. Partition $[a, b]$ into n equal subdivisions $\mathcal{P} : a = x_0 < x_1 < \cdots < x_n = b$, each of length $\Delta x = (b-a)/n$. Then the average of the values of f sampled at the right endpoint of each of these subintervals is

$$\frac{f(x_1) + f(x_2) + \cdots + f(x_n)}{n} = \frac{1}{b-a}\, [f(x_1) + f(x_2) + \cdots + f(x_n)]\, \Delta x,$$
$$(8.9)$$

a Riemann sum of $f/(b-a)$. Thus it is natural to define (8.8) as the average value of f. As an epigram, *the average value is the (signed) area over the base.*

Corollary. (mean value theorem for integrals) *A continuous function takes on its average value.* That is, if f is continuous on $[a, b]$, then for some $a < c < b$,

$$f(c) = \frac{1}{b-a} \int_a^b f(x)\, dx. \qquad (8.10)$$

Proof. Using a partition of one,

$$\inf_{[a,b]} f(x)(b-a) \;\leq\; \int_a^b f(x)\, dx \;\leq\; \sup_{[a,b]} f(x)(b-a).$$

But because f is continuous, it takes on its infimum and supremum as well as all values in between.

Remark C. Note that this mean value theorem (8.10) for integrals is our original mean value theorem (5.22) in disguise. For by setting

$$F(x) = \int_a^x f(t)\, dt,$$

we see that

$$F'(c) = \frac{F(b) - F(a)}{b-a}$$

is exactly (8.10).

Example 3. The average value of a linear function over an interval is its height at the midpoint:

$$\frac{1}{b-a} \int_a^b (mx + c)\, dx = \frac{1}{b-a} \left(\frac{mx^2}{2} + cx \right) \Bigg|_a^b$$

$$= \frac{1}{b-a} \left[\frac{mb^2 - ma^2}{2} + c(b - a) \right] = m\frac{a+b}{2} + c.$$

Example 4. The average value of $\sin x$ over a half-cycle is

$$\mathrm{av} = \frac{1}{\pi} \int_0^\pi \sin x\, dx = -\left. \frac{\cos x}{\pi} \right|_0^\pi = \frac{2}{\pi} \approx 0.637.$$

In contrast, the average over full cycles is of course 0.

8.4 Volumes

Result D. Suppose f is continuous on $[a, b]$. Let R be the region bounded by the graph of $y = f(x)$ and the x-axis from $x = a$ to $x = b$.

(a) When the region R is rotated about the x-axis, the resulting solid has volume

$$V = \pi \int_a^b f(x)^2\, dx. \qquad (8.11)$$

(b) Assume $0 < a < b$. When this same region R is instead rotated about the y-axis, the resulting volume is

$$V = 2\pi \int_a^b x|f(x)|\, dx. \qquad (8.12)$$

Verification. A Darboux sum verification is left as exercise 8.3. A verification by infinitesimals goes like this: Slice the solid of (a) with a plane perpendicular to the x-axis at x. The result is an infinitesimal cylinder of volume $dV = \pi r^2\, dx$ of radius $r = f(x)$ and height dx. Sum up these infinitesimal volumes.

In (b), this same vertical slice is rotated about the y-axis to produce a cylindrical shell of radius x, height $|y| = |f(x)|$, and infinitesimal thickness dx with volume $dV = 2\pi x|y|\, dx$.

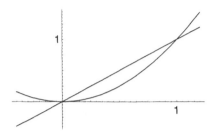

Figure 8.5 The region bounded by the curves
$y = x$ and $y = x^2$.

Example 5. Consider the (bounded) region R enclosed by $y = x$ and $y = x^2$, as seen in figure 8.5. When this region is rotated about the x-axis, the resulting volume is

$$V_a = \pi \int_0^1 x^2 - x^4 \, dx = \cdots = \frac{2\pi}{15}. \tag{8.13a}$$

When this same region is rotated about the y-axis, the resulting volume is

$$V_b = 2\pi \int_0^1 x(x - x^2) \, dx = \cdots = \pi/6. \tag{8.13b}$$

Note that this second result can also be obtained by slicing horizontally:

$$V_b = \pi \int_0^1 (\sqrt{y})^2 - y^2 \, dy = \cdots = \pi/6. \tag{8.13c}$$

8.5 Moments

Moments of a region or body are of central importance to mechanics and statistics. Visualize a collection of masses $m_1, m_2, \ldots, m_n$ at locations $p_1 = (x_1, y_1, z_1)$, $p_2 = (x_2, y_2, z_2), \ldots, p_n = (x_n, y_n, z_n)$, respectively. The *first moment* of these masses is the triple

$$M = \sum_{k=1}^n p_k \, m_k = \left(\sum_{k=1}^n x_k \, m_k, \sum_{k=1}^n y_k \, m_k, \sum_{k=1}^n z_k \, m_k \right). \tag{8.14}$$

Result E. The *center of mass O is the first moment divided by the total mass,* that is,

$$O = \frac{M}{m_1 + m_2 + \cdots + m_n}. \tag{8.15}$$

Verification. By definition, the center of mass is the origin of a coordinate system with respect to which the masses have first moment zero. Let $O = M/m = (\tilde{x}, \tilde{y}, \tilde{z})$ and $m = m_1 + \cdots + m_n$. Then

$$0 = M - mO = \sum_{k=1}^{n} m_k \cdot (x_k, y_k, z_k) - m_k \cdot (\tilde{x}, \tilde{y}, \tilde{z})$$

$$= \sum_{k=1}^{n} m_k \cdot [(x_k, y_k, z_k) - (\tilde{x}, \tilde{y}, \tilde{z})].$$

Thinking of a solid V as composed of infinitesimal masses dm at location $p = (x, y, z)$, one then leaps to the continuous version of (8.14) and (8.15): *The first moment of the solid V is the triple*

$$M = \int_V p \, dm \qquad (8.16a)$$

and its center of mass is at

$$O = (\tilde{x}, \tilde{y}, \tilde{z}) = \frac{M}{m}, \qquad (8.16b)$$

where m is the total mass

$$m = \int_V dm.$$

Example 6. Consider a uniform thin plate in the shape of the region in the plane bounded by $y = x^2$ and $y = 1$. See figure 8.6. Where does this plate balance?

Solution: We assume constant mass density per unit area. Clearly by symmetry, the geometric center *(centroid)* lies on the y-axis and so $\tilde{x} = 0$. As for its y-coordinate, consider the mass $dm = 2\sqrt{y} \, dy$

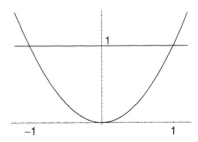

Figure 8.6 Where is the centroid of the region bounded by $y = 1$ and $y = x^2$?

of a infinitesimally thin horizontal slice with lever arm y and hence moment $dM = 2y\sqrt{y}\,dy$. Totalling all infinitesimal moments yields

$$\tilde{y} = \frac{\int_0^1 2y\sqrt{y}\,dy}{\int_0^1 2\sqrt{y}\,dy} = \frac{3}{5}.$$ (8.17a)

Alternatively, we may slice vertically, then concentrate all mass $dm = (1 - x^2)\,dx$ at the center of this infinitesimally thin strip with lever arm $y = (1 + x^2)/2$ to obtain

$$\tilde{y} = \frac{\int_{-1}^1 (1 + x^2)(1 - x^2)\,dx/2}{\int_{-1}^1 (1 - x^2)\,dx} = \frac{\int_0^1 (1 - x^4)\,dx}{2\int_0^1 (1 - x^2)\,dx} = \frac{3}{5}.$$ (8.17b)

A quantity central to modeling rotational motion is the *moment of inertia* of a collection of masses $m_1, m_2, \ldots, m_n$ *about an axis of rotation L.* Let r_k be the perpendicular distance of the mass m_k to the axis L. Then the moment of inertia about L is

$$I = \sum_{k=1}^n r_k^2\, m_k,$$ (8.18a)

with its continuous version

$$I = \int_V r^2\, dm.$$ (8.18b)

Example 7. What is the moment of inertia of a solid uniform cylinder about its axis?

Solution: Let the height of the cylinder be h and its radius a. Let us assume a mass density of δ per unit volume. We may collapse the mass of the cylinder onto one circular face to obtain a planar problem with mass density of $\sigma = h\delta$ per unit area. The mass at an infinitesimally thin annular ring of radius r from the axis is then $dm = 2\pi\sigma r\,dr$. Hence the moment of inertia is

$$I = \int_0^a r^2\, dm = 2\pi\sigma \int_0^a r^3\, dr = \frac{\pi\delta h a^4}{2}.$$ (8.19)

Result F. The *rotational kinetic energy* of a body rotating about a line L at angular velocity ω is

$$T = \frac{I\omega^2}{2}.$$ (8.20)

Thus *in rotational motion, the role of mass is played by the moment of inertia, velocity by angular velocity.*

Verification. The total kinetic energy T is the sum of all the kinetic energies of all the infinitesimal masses making up the solid V, giving

$$T = \int_V \frac{v^2\,dm}{2} = \int_V \frac{(r\omega)^2\,dm}{2} = \frac{\omega^2}{2}\int_V r^2\,dm = \frac{\omega^2 I}{2}.$$

8.6 Arclength

Result G. Suppose f is continuously differentiable on $[a, b]$. Then the arclength of the graph of $y = f(x)$ from $(a, f(a))$ to $(b, f(b))$ is

$$\Gamma = \int_a^b \sqrt{1 + f'(x)^2}\,dx. \tag{8.21}$$

Verification. Partition $[a, b]$ as usual: $a = x_0 < x_1 < \cdots < x_n = b$. The arclength Γ must be the supremum of the sum of all the straight-line distances

$$\sum_{i=1}^n \sqrt{(x_i - x_{i-1})^2 + (f(x_i) - f(x_{i-1}))^2}$$

$$= \sum_{i=1}^n \sqrt{1 + \left(\frac{f(x_i) - f(x_{i-1})}{x_i - x_{i-1}}\right)^2}\,\Delta x_i$$

$$= (\text{by MVT}) = \sum_{i=1}^n \sqrt{1 + f'(c_i)^2}\,\Delta x_i,$$

which are Riemann sums of the integral of (8.21).

Arguing instead with infinitesimals, the increment of arclength $ds = \sqrt{dx^2 + dy^2} = \sqrt{1 + (dy/dx)^2}\,dx$.

Example 8. Let us rederive the formula $C = 2\pi r$ for the circumference of a circle. Consider the half-circle of radius a given by $y = \sqrt{a^2 - x^2}$, $-a \le x \le a$. By (8.21), the arclength of the entire circle is (exercise 8.13)

$$\Gamma = 2\int_{-a}^a ds = 4\int_0^a ds = 4\int_0^a \sqrt{1 + y'^2}\,dx$$

$$= 4\int_0^a \sqrt{1 + \left(\frac{-x}{\sqrt{a^2 - x^2}}\right)^2}\,dx = 4a\int_0^a \frac{dx}{\sqrt{a^2 - x^2}}$$

$$= 4a\mathrm{Arcsin}\,(x/a)\Big|_0^a = 4a\mathrm{Arcsin}\,1 = 2\pi a. \tag{8.22}$$

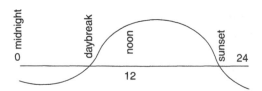

Figure 8.7 The rate of uptake of carbon dioxide is greater during photosynthesis than the rate of discharge at night. The net is the (signed) area under the curve, a positive uptake.

8.7 Accumulating Processes

The integral of a rate of production (or loss) of a quantity recovers the net accumulation of the quantity.

Example 9. Suppose $f(t)$ is the rate of CO_2 mass uptake of a given tree. During the day, sugars are formed from the ambient CO_2, light, and water. At night, some of these sugars are metabolized, releasing CO_2. The rate of intake and release of CO_2 follows a curve like that shown in figure 8.7. Thus,

$$U = \int_0^{24} f(t)\, dt$$

is the net mass of CO_2 locked away in the tissues of the tree per day.

8.8 Logarithms

Definition A. The *natural logarithm* is the function $\ln : (0, \infty) \longrightarrow \mathbf{R}$ given by the rule

$$\ln x = \int_1^x \frac{dt}{t}. \tag{8.23a}$$

Of course, because $1/x$ is continuous on $(0, \infty)$,

$$D \ln x = \frac{1}{x}. \tag{8.23b}$$

Note that $\ln 1 = 0$. Moreover, for any positive constant a,

$$D \left(\ln ax - \ln a - \ln x \right) = \frac{a}{ax} - \frac{1}{x} = 0,$$

and so $\ln ax = \ln a + \ln x + c$. But setting $x = 1$ reveals that $c = 0$. Thus we have the first of the laws of the logarithm:

$$\ln xy = \ln x + \ln y \tag{8.24}$$

for all positive x, y. In particular, for a natural number n,

$$\ln x^n = \ln (x \cdot x \cdots x) \quad (n \text{ times})$$
$$= \ln x + \ln x + \cdots + \ln x \quad (n \text{ times}) \; = n \ln x. \qquad (8.25)$$

Replacing x in (8.25) by $x^{1/n}$ yields that $(1/n) \ln x = \ln x^{1/n}$ and hence (exercise 8.15)

$$\ln x^r = r \ln x, \quad r \in \mathbf{Q}. \qquad (8.26)$$

Note that the logarithm is an increasing bijective continuous function from $(0, \infty)$ onto $(-\infty, \infty)$ (exercise 8.14). Therefore its inverse function $\ln^{-1} x = \exp x$ has by (5.26) the derivative $D \, \exp x = 1/(1/\exp x)$; that is,

$$D \, \exp x = \exp x. \qquad (8.27)$$

Note that $\exp 0 = 1$, since $\ln 1 = 0$ and that by (8.24) and (8.26),

$$\exp (x + y) = (\exp x) \cdot (\exp y), \quad x, y \in \mathbf{R} \qquad (8.28a)$$

and

$$\exp (rx) = (\exp x)^r, \quad x \in \mathbf{R}, \; r \in \mathbf{Q}. \qquad (8.28b)$$

Definition B. The real number

$$e = \exp (1) \qquad (8.29)$$

is called *Napier's natural base*. Its value is approximately 2.718281828 (exercise 8.21).

Note that for rational exponents, $e^r = (\exp 1)^r = \exp r$. But how can we make sense of e^π? Or $e^{\sqrt{2}}$?

Definition C. For any real number x, the meaning of the symbol e^x is given by

$$e^x = \exp x, \qquad (8.30a)$$

which is in complete agreement with the value e^r for r rational. For any positive b, we define

$$b^x = e^{x \ln b} = \exp (x \ln b). \qquad (8.30b)$$

A list of familiar results now follows easily (exercise 8.16).

Corollary A. (rules of exponents) For any positive b and real x, y,

$$b^{x+y} = b^x b^y, \quad (b^x)^y = b^{xy}, \quad b^{-x} = \frac{1}{b^x}, \quad b^0 = 1. \tag{8.31}$$

Corollary B. For $b > 0$, the function $f(x) = b^x$ is an everywhere differentiable bijective function from the reals to the positive reals with derivative

$$D\, b^x = b^x \ln b. \tag{8.32}$$

In particular,

$$D\, e^x = e^x. \tag{8.33}$$

8.9 Methods of Integration

There are two methods for integrating complicated integrals: *by substitution* and *by parts*.

Result H. (substitution) Suppose that $u = u(x)$ is continuously differentiable on $[a, b]$ and that f is continuous on at least the interval $u([a, b])$. Then

$$\int_a^b f(u(x))u'(x)\, dx = \int_{u(a)}^{u(b)} f(u)\, du. \tag{8.34}$$

Proof. By (7.25), the derivatives with respect to b of both sides of (8.34) agree, thus they differ by a constant. But setting $b = a$ reveals the constant to be zero.

Example 10. Make the substitution $u = 1 + x^2$ so that

$$\int_0^1 x(1 + x^2)^5\, dx = \int_1^2 u^5\, du/2 = \left.\frac{u^6}{12}\right|_{u=1}^{u=2}.$$

Example 11. Often the formula of substitution (8.34) is run from right to left. For example, let us rederive the formula $A = \pi r^2$ for the area of a disk. We employ the substitution $u = r \sin\theta$ whereupon $du = r \cos\theta\, d\theta$. Hence

$$A = 4\int_0^r \sqrt{r^2 - u^2}\, du = 4\int_0^{\pi/2} \sqrt{r^2 - r^2 \sin^2\theta}\, r \cos\theta\, d\theta$$

$$= 4r^2 \int_0^{\pi/2} \cos^2\theta\, d\theta = 4r^2 \int_0^{\pi/2} \frac{1 + \cos 2\theta}{2}\, d\theta = \pi r^2.$$

Result I. (integration by parts[2]) Suppose u and v are both continuously differentiable on $[a, b]$. Then

$$\int_a^b u\, dv = uv \Big|_a^b - \int_a^b v\, du,$$

that is,

$$\int_a^b u(x)v'(x)\, dx = u(x)v(x) \Big|_a^b - \int_a^b u'(x)v(x)\, dx. \qquad (8.35)$$

Proof. Merely integrate the product rule $(uv)' = uv' + u'v$.

Example 12. By setting $u = x$ and $dv = e^x\, dx$ we obtain

$$\int_0^1 xe^x\, dx = xe^x \Big|_0^1 - \int_0^1 e^x\, dx = e - (e - 1) = 1.$$

Example 13. The method of *partial fractions* is the frequency-domain equivalent of integration by parts—see [MacCluer, 2000]. It is used to integrate rational functions. For instance,

$$\int_0^1 \frac{x}{x^2 + 3x + 2}\, dx = \int_0^1 \left(\frac{A}{x+1} + \frac{B}{x+2} \right) dx$$

$$= \int_0^1 \left(\frac{-1}{x+1} + \frac{2}{x+2} \right) dx = \ln \frac{(x+2)^2}{x+1} \Big|_0^1.$$

8.10 Improper Integrals

Let us now extend the concept of the integral to unbounded integrands and intervals.

Example 14. (unbounded integrand) Consider the integral

$$I = \int_0^1 \frac{dx}{\sqrt{x}}. \qquad (8.36)$$

Note that although the interval of integration is bounded, the integrand $f(x) = 1/\sqrt{x}$ is unbounded at $x = 0$ and so the hypotheses for the Riemann integral are not satisfied.

---

[2]Not only is integration by parts a useful computational trick, but it crops up in important mathematics—for example, it is crucial in prime number estimates and in the modern Sobolev school of partial differential equations.

However, also note that f is bounded on every subinterval $[\epsilon, b]$, $\epsilon > 0$, and in fact

$$I_\epsilon = \int_\epsilon^1 \frac{dx}{\sqrt{x}} = 2\sqrt{x}\,\Big|_\epsilon^1 = 2 - 2\sqrt{\epsilon},$$

and so in the limit, as $\epsilon \to 0$, it would be natural to claim $I = 2$. Thus (8.36) is an example of a *convergent improper integral*.

Definition D. Suppose f is Riemann integrable on every closed subinterval of (a, b). Then the (possibly improper) integral

$$\int_a^b f(x)\, dx$$

is said to *converge to the value I* when

$$\lim_{\delta \to 0^+} \lim_{\epsilon \to 0^+} \int_{a+\epsilon}^{b-\delta} f(x)\, dx = \lim_{\epsilon \to 0^+} \lim_{\delta \to 0^+} \int_{a+\epsilon}^{b-\delta} f(x)\, dx = I, \qquad (8.37)$$

and we extend the meaning of equality by writing

$$\int_a^b f(x)\, dx = I.$$

Example 15. (unbounded interval) Consider the improper integral

$$I = \int_1^\infty \frac{dx}{x^2}. \qquad (8.38)$$

Note that

$$\lim_{N \to \infty} \int_1^N \frac{dx}{x^2} = -\lim_{N \to \infty} \frac{1}{x}\Big|_1^N = 1$$

and so (8.38) is *convergent to the value 1*.

Definition E. Suppose f is Riemann integrable on every closed and bounded subinterval of $[a, \infty)$. Then the improper integral

$$\int_a^\infty f(x)\, dx$$

is said to *converge to the value I* when

$$\lim_{N \to \infty} \int_a^N f(x)\, dx = I \qquad (8.39)$$

and we extend the meaning of equality by writing

$$\int_a^\infty f(x)\,dx = I.$$

(There is the obvious similar notion when the lower limit of integration is $-\infty$.)

Occasionally integrals are improper for several reasons, for example

$$I = \int_0^\infty f(x)\,dx = \int_0^\infty \frac{dx}{\sqrt{x}\sqrt{|1 - x^2|}}.$$

In such cases, merely divide the interval of integration and deal with each impropriety separately. In this example, rewrite the integral as

$$I = \int_0^{1/2} f(x)\,dx + \int_{1/2}^1 f(x)\,dx + \int_1^2 f(x)\,dx + \int_2^\infty f(x)\,dx. \qquad (8.40)$$

If each summand converges (exercise 8.26), one says the the original improper integral *converges*. If any one summand diverges, so does the original integral.

8.11 Statistics

Statistics is a profound example of the "unreasonable efficacy of mathematics." Although a given phenomenon may be fundamentally random, it may nevertheless exhibit great regularity on average. The probability (frequency) that a certain measurement X of the phenomenon has value between a and b can often be predicted as an integral of the form

$$\text{prob}(a \le X \le b) = \int_a^b f(x)\,dx, \qquad (8.41a)$$

where f is called the *probability density function* of the measurement X. Of course,

$$\text{prob}(-\infty < X < \infty) = \int_{-\infty}^\infty f(x)\,dx = 1. \qquad (8.41b)$$

The expected value of the outcome *(mean)* is the first moment

$$\mu = \int_{-\infty}^\infty x f(x)\,dx, \qquad (8.42a)$$

which, because the total mass is 1, is the the x-coordinate of the center of mass of the distribution. The *variance* v is the moment of inertia about the mean:

$$v = \sigma^2 = \int_{-\infty}^{\infty} (x - \mu)^2 f(x) \, dx, \qquad (8.42b)$$

where $\sigma > 0$ is called the *standard deviation*. The smaller the standard deviation, the more likely the outcomes cluster about the mean.

Example 16. Take a stopwatch to a supermarket and record the time between arrivals at a checkout lane. You will find close agreement with *Poisson flow:* The probability that the time T before the next arrival is at most t_0 is given by

$$\text{prob}(T \leq t_0) = \frac{1}{a} \int_0^{t_0} e^{-t/a} \, dt,$$

where a is the mean time between arrivals. So the statistical outcomes for this experiment (of measuring time between arrivals) are modeled by the probability density function

$$f(t) = \begin{cases} (1/a)e^{-t/a} & \text{if } 0 \leq t \\ 0 & \text{if } t < 0. \end{cases}$$

Example 17. A certain scholarship aptitude test is designed under the belief that scores should be *normally distributed* with mean $\mu = 500$ and standard deviation $\sigma = 100$. So the probability that a randomly chosen student's score X will fall between (say) $a = 550$ and $b = 675$ is given by

$$\text{prob}(a \leq X \leq b) = \frac{1}{\sqrt{2\pi}\,\sigma} \int_a^b e^{\frac{(x-\mu)^2}{2\sigma^2}} \, dx. \qquad (8.43)$$

This famous integral has no elementary antiderivative and must be calculated numerically. See §8.13 and exercise 8.76.

Any experimental outcome that is the superposition of many unrelated phenomena will have this famous *Gaussian* (normal) distribution (8.43).[3]

---

[3]This is an informal statement of the *central limit theorem* of statistics to be found in any text on statistics.

8.12 Quantum Mechanics

The rules of mechanics change at the atomic scale—they become statistical rather than deterministic. One can obtain only expected (average) measurements. Mechanical rules are no longer relationships between measured quantities; they become instead the analogous relationships between the instruments taking the measurements.

The most that can be known about a particle is its *wave function* ψ, where $f(x) = |\psi(x)|^2$ is a probability density function. The wave function ψ is a solution of *Schrödinger's equation*

$$i\hbar\dot{\psi} = H\psi, \tag{8.44}$$

where H is the mathematical analog of the instrument measuring total energy. When any one measurement is taken, the wave function ψ "collapses" to one of its *stationary states,* that is, a time-independent solution of

$$H\psi = E\psi, \tag{8.45}$$

where E is the energy of that state. The mathematics necessary to model all this is called *functional analysis,* a sort of infinite dimensional matrix theory. To learn more about quantum mechanics, see the delightful book by [Davies] and my own view [MacCluer, 2004]. But let us sample the flavor of the subject by working through one problem to see how the integral comes into play.

Example 18. A particle of mass m is trapped in the interval $[0, \pi]$ by an infinite external potential. There is 0 potential within the interval. Thus all the particle's energy is kinetic. Schrödinger's equation (8.45) for the stationary states then becomes

$$-\frac{\hbar^2}{2m}\psi'' = E\psi; \tag{8.46}$$

that is, the wave function ψ satisfies the ODE of harmonic motion $\psi'' + \omega^2\psi = 0$, where in this case $\omega = \sqrt{2mE}/\hbar$. The solutions to the harmonic motion ODE are the sinusoidals $\psi(x) = a\sin(\omega x - \theta)$. But because the particle is trapped, $\psi(x)$ vanishes outside $(0, \pi)$. Thus $\theta = 0$ and $\omega = n$, $n \in \mathbf{N}$. Because $|\psi|^2$ is a probability distribution,

$$1 = \int_{-\infty}^{\infty} |\psi(x)|^2\,dx = \int_{0}^{\pi} |\psi(x)|^2\,dx$$

$$= a^2 \int_{0}^{\pi} \sin^2 nx\,dx = \frac{a^2}{2} \int_{0}^{\pi} (1 - \cos 2nx)\,dx = \pi a^2/2,$$

hence $a = \sqrt{2/\pi}$. Thus the wave function of the n-th stationary state is

$$\psi_n(x) = \sqrt{\frac{2}{\pi}} \sin nx, \tag{8.47a}$$

that must because of (8.46) have (kinetic) energy

$$E_n = \frac{\hbar^2 n^2}{2m}, \quad n \in \mathbf{N}. \tag{8.47b}$$

It so happens that position (displacement) is measured by mutiplication by x. Thus the expected location of the particle is

$$\langle Q\psi, \psi \rangle = \int_0^\pi x|\psi(x)|^2 \, dx$$
$$= \frac{2}{\pi} \int_0^\pi x \sin^2 nx \, dx = \text{(exercise 8.28)} = \pi/2. \tag{8.48}$$

Thus on average, the particle is to be found at the midpoint of the interval.

8.13 Numerical Integration

Although extensive tables of antiderivatives exist [CRC], many important integrals cannot be given in terms of elementary functions. One way to think about the history of mathematics is that when mathematicians encounter an equation with a solution inexpressible in terms of known quantities, the solution is named, tabulated, and added to the list of primitive notions.

For example, an integral important to statistics and heat transfer is the *error function*

$$\mathrm{erf}(x) = \frac{2}{\sqrt{\pi}} \int_0^x e^{-\beta^2} \, d\beta, \tag{8.49}$$

an integral with no antiderivative in terms of elementary functions. Before the availability of cheap computing, such integrals were often tabulated by analytic techniques, such as Taylor expansions, as follows: Since

$$e^x = 1 + x + \frac{x^2}{2!} + \frac{x^3}{3!} + \cdots,$$
$$\int_0^x e^{-\beta^2} \, d\beta = \int_0^x \left(1 - \beta^2 + \frac{\beta^4}{2!} - \frac{\beta^6}{3!} + \cdots\right) d\beta.$$

Nowadays we employ the digital computer to compute cleverly chosen Riemann sums.

Objective. To estimate the value of the Riemann integral

$$I = \int_a^b f(x) \, dx. \tag{8.50}$$

Approach. Partition the interval $[a, b]$ into n equal parts of length $\Delta x = (b-a)/n : a = x_0 < x_1 < x_2 < \cdots < x_n = b$, where $x_i = a + i \cdot \Delta x$.

The five standard algorithms. The Riemann sums

$$L_n = \Delta x \cdot (f(x_0) + f(x_1) + \cdots + f(x_{n-1})), \tag{8.51a}$$

$$R_n = \Delta x \cdot (f(x_1) + f(x_2) + \cdots + f(x_n)), \tag{8.51b}$$

$$M_n = \Delta x \cdot \left(f\left(\frac{x_0 + x_1}{2}\right) + f\left(\frac{x_1 + x_2}{2}\right) \right.$$
$$\left. + \cdots + f\left(\frac{x_{n-1} + x_n}{2}\right) \right) \tag{8.51c}$$

are called the *left-hand, right-hand,* and *midpoint rule* estimates for (8.50), respectively.

The estimates

$$T_n = \frac{L_n + R_n}{2} = \Delta x \cdot \left(\frac{f(x_0)}{2} + f(x_1) + f(x_2) + \cdots + f(x_{n-1}) + \frac{f(x_n)}{2} \right) \tag{8.52}$$

and (when n is even),

$$S_n = \frac{\Delta x}{3} \cdot (f(x_0) + 4f(x_1) + 2f(x_2) + 4f(x_3) + \cdots + 4f(x_{n-1}) + f(x_n)) \tag{8.53}$$

are called the *trapezoid* and *Simpson rule* estimates for (8.50), respectively.

Example 19. Let us estimate π using each of these five rules and the integral

$$\frac{\pi}{4} = \text{Arctan } 1 = \int_0^1 \frac{dx}{1 + x^2}. \tag{8.54}$$

Partition the interval $[0, 1]$ into $n = 4$ equal parts: $0 < 1/4 < 1/2 < 3/4 < 1$. Then the five estimates to $\pi/4 \approx 0.785398164$ are

$$L_4 = \frac{1}{4}\left(\frac{1}{1+0^2} + \frac{1}{1+(1/4)^2} + \frac{1}{1+(1/2)^2} + \frac{1}{1+(3/4)^2}\right) \approx 0.84529,$$

$$R_4 = \frac{1}{4}\left(\frac{1}{1+(1/4)^2} + \frac{1}{1+(1/2)^2} + \frac{1}{1+(3/4)^2} + \frac{1}{1+1^2}+\right) \approx 0.72029,$$

$$M_4 = \frac{1}{4}\left(\frac{1}{1+(1/8)^2} + \frac{1}{1+(3/8)^2} + \frac{1}{1+(5/8)^2} + \frac{1}{1+(7/8)^2}\right) \approx 0.78670,$$

$$T_4 = \frac{1}{4}\left(\frac{1}{1+0^2} + \frac{2}{1+(1/4)^2} + \frac{2}{1+(1/2)^2} + \frac{2}{1+(3/4)^2} + \frac{1}{1+1^2}\right) \approx 0.78229,$$

$$S_4 = \frac{1}{12}\left(\frac{1}{1+0^2} + \frac{4}{1+(1/4)^2} + \frac{2}{1+(1/2)^2} + \frac{4}{1+(3/4)^2}\frac{1}{1+1^2}\right) \approx 0.785392.$$

Notice the stunning accuracy of Simpson's rule.

It is easy to instruct machines to perform these numerical integrations:

Pseudocode. (for approximating (8.50) using n subdivisions)

```
n = 4
x = a
dx = (b-a)/n
L = 0, R = 0, M = 0
T = f(x), S = f(x)
c = 1
loop (n times)
   x = x + dx
   L = L + f(x-dx)
   R = R + f(x)
   M = M + f(x-dx/2)
   T = T + 2f(x)
   S = (3 + c)f(x)
   c = -c
return to loop
L = L dx, R = R dx, M = M dx, T = (T - f(x)) dx/2
S = (S - f(x)) dx/3
```

Remark D. It is clear why the lefthand, righthand, midpoint, and trapezoid rules estimate the integral, since they are either Riemann sums or averages of Riemann sums. But why does Simpson's rule work and why so well? It works because (exercise 8.29)

$$S_n = \frac{L_n + R_n + M_{n/2}}{3} \tag{8.55}$$

and thus S_n is also an average of Riemann sums. But why is it so accurate? It is because Simpson's rule replaces the integrand with piecewise quadratics rather than piecewise linear elements. This yields an error that decreases with the *fourth* power of the number of subdivisions.

Result J. Assume the integrand f of (8.50) has at least its first four derivatives continuous on $[a, b]$. Let

$$m = \sup_{[a,b]} |f^{(4)}(x)|.$$

Then for a partition of $[a, b]$ into n (an even number) equal subintervals of length $\Delta x = (b - a)/n$, the absolute Simpson error

$$\left| S_n - \int_a^b f(x)\, dx \right| \le \frac{(b - a)^5 m}{180 n^4}. \tag{8.56}$$

Proof. Exercise 8.31.

Example 20. How many subdivisions n are necessary to achieve eight-place accuracy in the Simpson estimate of

$$\ln 2 = \int_1^2 \frac{dx}{x}? \tag{8.57}$$

Solution: By (8.56) we must choose n even and large so that

$$\frac{(2 - 1)^5 m}{180 n^4} < 5 \times 10^{-9},$$

where m is the maximum value on $[1, 2]$ of $|(1/x)''''| = |4!/x^5|$, i.e., $m = 24$. Thus it is sufficient that

$$n > \left(\frac{24 \cdot 10^9}{5 \cdot 180} \right)^{1/4} \approx 71.9.$$

Hence seventy-two subdivisions suffice.

Exercises

8.1   Compute the work needed to move a particle from $x = 1$ to $x = 4$ against the force field $f(x) = -1/x^2$.
        Answer: 3/4.

8.2   Compute the work needed to compress a spring with spring constant
       $k = 3$ N/m and natural length $L_N = 2$ m from length $L = 1$ m to
       $L = 4$ m.
           Answer: 9/2 N-m.

8.3   Using Darboux sums, provide a verification of result D.

8.4   Deduce from example 8 the formula $A = \pi r^2$ for the area of a disc of
       radius r. Deduce in turn the formula $V = 4\pi r^3/3$ for the volume V of
       a ball of radius r from its surface area $S = 4\pi r^2$.

8.5   Compute the average value of $\sqrt{x}$ over $[0, 2]$.
           Answer: $2\sqrt{2}/3$.

8.6   A sinusoidal voltage $v = a \sin \omega t$ is applied to a resistor of resistance R.
       Show that the energy (in Joules) consumed by the resistor in heat
       every complete period T is $a^2 T/2R$. Deduce that an alternating
       voltage of peak value a can only deliver the energy of the lower DC
       voltage $a/\sqrt{2}$. (For example, the nominal "120 volt AC" North
       American house voltage actually peaks at approximately 170 volts.)
           Outline: Instantaneous power P (in Watts) through the resistor is by
           Ohm's law, $P = vi = v^2/R = (a^2/R) \sin^2 \omega t$. Energy is the time-
           integral of power.

8.7   Verify the formulas (8.13).

8.8   Suppose we have n masses m_i moving in empty space, respectively
       located at $(x_i(t), y_i(t), z_i(t))$ at time t. Let $O(t)$ be the location of the
       center of mass of these n masses at time t. Prove that O is moving
       along a straight line at constant speed.
           Hint: Differentiate the coordinates of $O(t)$ and use that inertial
           forces sum to 0.

8.9   Verify (8.17).

8.10   Calculate the moment of inertia of a solid uniform ball of density δ
       and radius a.
           Answer: $I = 8\pi \delta a^5/15$.
           Outline: Consider cylindrical shells about the axis of rotation,

$$I = \int_0^a r^2 dm = 2\delta \int_0^a r^2 2\pi r \sqrt{a^2 - r^2}\, dr.$$

       Employ the substitution $r = a \sin \theta$.

8.11   Verify *Pappus's First Law:* The volume V obtained by revolving a planar region on one side of a line L about L is the area A of the region multiplied by the circumference traveled by the centroid of the region: $V = 2\pi \bar{r} A$. Think of a donut.

8.12   Consider a curve in the plane given by $y = f(x)$ for $a \le x \le b$. Argue that when this curve is rotated about the x-axis, the resulting surface area generated is

$$S = 2\pi \int_a^b y \, ds = 2\pi \int_a^b f(x) \sqrt{1 + f'(x)^2} \, dx.$$

Deduce *Pappus's Second Law:* The surface area S obtained by rotating a plane curve on one side of a line about this line is the product of the distance $2\pi r$ traveled by the geometric centroid[4] of the curve with the arclength Γ of the curve: $S = 2\pi r \Gamma$.
    For example, the (slanted) surface area of a right circular cone of radius r is its slant height times πr.

8.13   Verify all steps leading to (8.22).
       Outline: This is more delicate than it first appears. Note that the integrand is unbounded as $x \to a$. It is an *improper integral* that nevertheless converges. Integrate up to $a - \epsilon$, then let ϵ go to 0.

8.14   Show from the definition (8.23) that logarithm is a bijective continuous function from $(0, \infty)$ onto $(-\infty, \infty)$.

8.15   Verify (8.26).

8.16   Prove the rules of exponents (9.31).

8.17   Prove the differentiation formula $D\, b^x = b^x \ln b$ from (8.30b).

8.18   Estimate $\ln 2$ and $\ln 3$ using lower and upper sums of (8.23a), respectively. Deduce that $2 < e < 3$.

8.19   Deduce from (8.30b) that $\alpha \ln x = \ln x^\alpha$ for any real α and $x > 0$.

---

[4]The centroid of a curve is the center of mass were the curve to be made from a uniform wire.

8.20   For $b > 0$, the *logarithm to base b* of x, in symbols $g(x) = \log_b x$, is defined as the inverse function of $f(x) = b^x$. Prove that for $x > 0$,

$$\log_b x = \frac{\ln x}{\ln b}.$$

Deduce that $\log_b (xy) = \log_b x + \log_b y, \quad \alpha \log_b x = \log_b x^\alpha$, and

$$D \log_b x = \frac{1}{x \ln b}.$$

8.21   Estimate e to six places.
        Outline: Employing Taylor's theorem (exercise 5.20),

$$e = 1 + \frac{1}{1!} + \frac{1}{2!} + \frac{1}{3!} + \cdots + \frac{1}{n!} + \frac{e^c}{(n+1)!}.$$

But $e^c < e^1 < 3$. Thus choose n large enough so that

$$\frac{3}{(n+1)!} < 5 \times 10^{-7}.$$

8.22   Let $b > 0$. Prove that there is at most one extension of $f(r) = b^r$ on $\mathbf{Q}$ to a continuous function on $\mathbf{R}$.
        Outline: Show that if f is uniformly continuous on the rationals in $[-n, n]$, then there is but one continuous extension to $[-n, n]$.

8.23   Compute

$$\int_0^1 \frac{x^2 \, dx}{1 + x^6}.$$

8.24   For what p does the improper integral

$$\int_0^1 \frac{dx}{x^p}$$

converge?
        Answer: $p < 1$.

8.25   For what p does the improper integral

$$\int_1^\infty \frac{dx}{x^p}$$

converge?
        Answer: $p > 1$.

8.26   Show that each improper summand of (8.40) converges.
        Hint: Use upper estimates.

8.27   Prove that

$$\int_0^\infty \frac{\sin x}{x}\, dx$$

converges (to in fact $\pi/2$).

   Outline:

$$\int_0^\infty \frac{\sin x}{x}\, dx = \int_0^\pi \frac{\sin x}{x}\, dx + \int_\pi^\infty \frac{\sin x}{x}\, dx.$$

   Integrate the second integral on the right by parts and apply the comparison theorem (exercise 8.49).

8.28   Verify (8.48).

8.29   Verify (8.55). That is, prove that the Simpson's rule approximation is the average of three other approximations:

$$S_n = \frac{L_n + R_n + M_{n/2}}{3}.$$

8.30   **(Project)** Use the inverse function of

$$I(y) = \int_0^y \frac{dx}{\sqrt{1 - x^2}},$$

   (extended periodically) as an analytic definition of $\sin x$. Derive all trigonometric results of §5.5 without dubious appeals to geometry.

8.31* Prove the error formula (8.56) for Simpson's rule.

   Outline: Reduce to the case of two subdivisions then set

$$E(h) = \int_{-h}^h f(x)\, dx - \frac{h}{3}[f(-h) + 4f(0) + f(-h)].$$

   Note that $0 = E(0) = E'(0) = E''(0) = E'''(0)$, and so by the Taylor formula with integral remainder,

$$E(h) = \frac{1}{3!}\int_0^h E^{(4)}(x)(h - x)^3\, dx = \frac{E(c)}{3!} = \cdots$$

8.32   Estimate the average value of the function f on $[0, 4]$ knowing only the sampled data

x	0	1	2	3	4
$f(x)$	1.2	2.3	2.9	3.1	2.8

   Recommended engineering practice: Use Simpson's rule.

8.33   Verify *Cavalieri's First Rule:* The volume of a generalized cylinder is the product of the area of its base with its altitude: $V = Ah$.

(A generalized cylinder is formed by vertically lifting a horizontal planar region parallel with itself.)

Hint: Slice parallel to the base.

8.34   Verify *Cavalieri's Second Rule:* The volume of a generalized cone is one third of the product of the area of its base with its altitude: $V = Ah/3$.

(A generalized cone is formed by all line segments connecting a point P to all points of a planar (base) region.)

Hint: Slice parallel to the base and use that

$$\int_0^h x^2 \, dx = \frac{h^3}{3}.$$

8.35   Use Pappus's rules to deduce the surface area and volume of a ball from the circumference and area of a disk. What is the surface area and volume of a donut?

8.36   The *Gamma function* is given by the improper integral

$$\Gamma(s) = \int_0^\infty x^{s-1} e^{-x} \, dx, \quad s > 0.$$

Show that $\Gamma(1) = 1$ and that $\Gamma(s+1) = s\Gamma(s)$. Deduce that $\Gamma(n+1) = n!$.

Remark: Many calculators have the Gamma function built in disguised as $\Gamma(s) = (s-1)!$. Experiment with yours. Does entering "$(-0.5)!$" yield $\sqrt{\pi} = \Gamma(1/2) \approx 1.77245$? If so, graph $y = \Gamma(x)$.

8.37   The *Beta function B* is defined by the integral

$$B(p, q) = \int_0^1 x^{p-1}(1 - x)^{q-1} \, dx \quad p, q > 0.$$

Prove by a clever change of variables that for $\alpha, \beta > -1$,

$$\int_0^{\pi/2} \cos^\alpha \theta \sin^\beta \theta \, d\theta = \frac{1}{2} B\left(\frac{\alpha+1}{2}, \frac{\beta+1}{2}\right).$$

8.38   Calculate the volume $V_n(r)$ of the first several n-dimensional balls $x_1^2 + x_2^2 + \cdots + x_n^2 \le r^2$.

Answer: $V_1(r) = 2r$, $V_2(r) = \pi r^2$, $V_3(r) = 4\pi r^3/3$, $V_4(r) = \pi^2 r^4/2$, $V_5(r) = 8\pi^2 r^5/15$.

Hint: By slicing, the n-ball of radius r has volume

$$V_n(r) = \int_{-r}^{r} V_{n-1}(\sqrt{r^2 - x^2})\, dx.$$

8.39   Calculate the work done against the force $f(x) = -x^3 e^{-x}$ while moving from $x = 1$ to $x = 2$.

8.40   Derive the integration formula

$$\int_a^b e^{\beta x} \sin x\, dx = \frac{\beta \sin x - \cos x}{1 + \beta^2} e^{\beta x} \Big|_a^b.$$

Hint: Integrate twice by parts.

8.41   Integrate

$$\int_a^b \mathrm{Arctan}\, x\, dx.$$

Hint: Integrate by parts with $u = \mathrm{Arctan}\, x$.

8.42   Integrate

$$\int_a^b \ln x\, dx.$$

8.43   Integrate

$$\int_a^b \frac{x^2 - 3x - 1}{(x-1)(x-2)(x^2+x+1)}\, dx.$$

Suggestion: Employ the partial fraction expansion

$$\frac{x^2 - 3x - 1}{(x-1)(x-2)(x^2+x+1)} = \frac{A}{x-1} + \frac{B}{x-2} + \frac{Cx+D}{x^2+x+1}.$$

8.44   Find the centroid of the planar region bounded above by $y = \sin x$ and below by the x-axis, $0 \le x \le \pi/2$.

8.45   Suppose the ground state (lowest energy stationary state) of a particle trapped in $[0, \infty)$ has wave function $\psi(x)$, where $|\psi(x)|^2 = xe^{-x}$. Find the expected location of this particle.

Outline: Since the instrument Q observing displacement is given by the rule $Q\psi = x\psi$,

$$\langle Q\psi, \psi \rangle = \langle x\psi, \psi \rangle = \int_0^{\infty} x^2 e^{-x}\, dx = \cdots = 2.$$

8.46   The *Rayleigh* probability density function

$$f(r) = \frac{re^{-r^2/2\sigma^2}}{\sigma^2}$$

is a good model (for instance) of the distance error r from the
bullseye in target shooting (since horizontal and vertical error are
normally distributed). What is the mean error?
    Answer:

$$\mu = \sqrt{2}\sigma\Gamma\left(\frac{3}{2}\right) = \sigma\sqrt{\frac{\pi}{2}}.$$

8.47   Estimate the arclength of $y = \sin x$ for $0 \le x \le \pi$.
(The integral for arclength in this case is *elliptic* and hence has no
antiderivative in terms of elementary functions. Numerical methods
are necessary.)

8.48   What is the resulting surface area S when the curve
$y = 1 - x^2$, $0 \le x \le 2$, is rotated about the y-axis?
    Outline: By exercise 8.12,

$$S = 2\pi\int_0^2 x\,ds = \cdots = \frac{\pi}{6}[(17)^{3/2} - 1].$$

8.49   Prove the *comparison test:* Suppose $0 \le f(x) \le g(x)$ on $[a, \infty)$. Then

$$\int_a^\infty g(x)\,dx \text{ convergent implies } \int_a^\infty f(x)\,dx \text{ is convergent.}$$

8.50   Using the comparison test of exercise 8.49, prove the convergence of

$$\int_0^\infty \frac{dx}{3x^2 + 2x + 2 + \sin x + \sqrt{x}}.$$

8.51   Calculate

$$\int_2^\infty \frac{dx}{x\ln^2 x}.$$

8.52   Using Simpson's rule with four subdivisions, estimate

$$I = \int_0^1 \frac{dx}{1 + x^3}.$$

8.53   How many subdivisions are sufficient to approximate

$$I = \int_0^1 e^{x^2}\,dx$$

to three places via Simpson's rule?

8.54   Develop an integration formula for integrals of type

$$\int x^n \sin \alpha x \, dx.$$

8.55   Find the following antiderivatives:

a. $\int \sqrt{1 + \cos 2x} \, dx$

b. $\int \dfrac{dx}{\sqrt{e^{2x} - 1}}$

c. $\int \dfrac{dx}{1 + \sqrt{x}}$

d. $\int \dfrac{x^3 \, dx}{x^2 - 2x + 1}$

e. $\int \dfrac{x^2 \, dx}{x^4 - 16}$

f. $\int \dfrac{4x \, dx}{x^4 + 1}$

g. $\int \tan^2 x \, dx$

h. $\int \cos^3 x \sin^2 x \, dx$

8.56   The *hyperbolic trigonometric* functions are defined by

$$\cosh x = \frac{e^x + e^{-x}}{2} \quad \text{and} \quad \sinh x = \frac{e^x - e^{-x}}{2}.$$

Show that every point on the hyperbola $x^2 - y^2 = 1$ is of the form $(\pm \cosh t, \pm \sinh t)$.

8.57   Prove the differentiation formulas

$$D \cosh x = \sinh x,$$
$$D \sinh x = \cosh x,$$

and derive the Taylor expansions

$$\cosh x = 1 + \frac{x^2}{2!} + \frac{x^4}{4!} + \cdots,$$
$$\sinh x = x + \frac{x^3}{3!} + \frac{x^5}{5!} + \cdots.$$

8.58   Argue that the complex exponential $e^{i\theta}$, if defineable at all, must satisfy $e^{i\theta} = \cos \theta + i \sin \theta$.
Hint: Think Taylor series.

8.59   Galileo estimated the acceleration g of gravity by rolling solid balls down an inclined plane at extreme inclinations. Show that because of rotational inertia, his estimates of g were doomed to be $5g/7$.

   Hint: The kinetic energy gained is the potential energy lost. Total kinetic energy is the sum of translational and rotational kinetic energies. Apply exercise 8.10.

8.60   A hoop, a solid cylinder, and a solid ball are simultaneously allowed to roll down an inclined ramp. In what order do they finish at the bottom?

   Answer: The ball is first, then cylinder, then hoop.

8.61   The *convolution* of two continuous functions f, g on $[0, \infty)$ is the (continuous) function $h = f * g$ given by the rule

$$h(t) = \int_0^t f(\alpha) g(t - \alpha) \, d\alpha.$$

   Calculate the convolutions $t * t$ and $t * e^{-t}$.
   Answers: $t^3/6$ and $-1 + t + e^{-t}$.

8.62   The *Laplace transform* F of a continuous function f on $[0, \infty)$ is (where convergent) given by the improper integral

$$F(s) = \int_0^\infty f(t) e^{-st} \, dt.$$

   Show that the Laplace transform of $f(t) = t^n$ is $F(s) = n!/s^{n+1}$ for $s > 0$ and that the transform of $g(t) = e^{-at}$ is $G(s) = 1/(s + a)$ for $s > -a$.

8.63   Ignoring technical convergence details, substantiate that the Laplace transform of a convolution is the product of their transforms. Using this result, redo exercise 8.61. Notice how integration by parts in the time domain becomes partial fraction expansion in the Laplace s-domain.

8.64   Show that the work V required to move a satellite of mass m at a distance r from a planet of mass M to infinity is $V = GMm/r$.

8.65   Compute the energy (work) W necessary to lift a satellite of mass m from the surface vertically to a distance r from the center of the Earth.
   Answer: $W = gmR(r - R)/r$, where R is the radius of the Earth and g is the acceleration at its surface.

8.66   (Coulomb's Law) It is observed that two charges Q_1 and Q_2 of the same polarity will repel one another with the force (in Newtons)

$$F = \frac{1}{4\pi\epsilon_0}\frac{Q_1 Q_2}{r^2},$$

where r is the distance (in meters) separating the two charges and where ϵ_0 is the *permittivity* of free space (in the appropriate units). Glue a $Q_1 = 2$ Coulomb charge to the origin $x = 0$. Calculate the work required to move a $Q_2 = 6$ Coulomb charge from $x = 4$ to $x = 1$. *Answer:* $9/4\pi\epsilon_0$ N-m.

8.67   Suppose the z-axis is uniformly charged at $\rho > 0$ Coulomb per meter. Show that the electrostatic force F on a unit charge of $Q = 1$ Coulomb located at (x, y, z) is directed perpendicular to the z-axis and is of magnitude

$$F = \frac{\rho Q}{2\pi\epsilon_0\sqrt{x^2 + y^2}}.$$

8.68   (Cauchy principal value) How far should we push the enlargement of equality? Consider the following ridiculous result when the limits at infinity are taken simultaneously:

$$P.V. \int_{-\infty}^{\infty} x^2 \sin x \, dx = \lim_{N\to\infty} \int_{-N}^{N} x^2 \sin x \, dx = 0.$$

Should we allow this even more forgiving improper integral?

8.69   **(Project)** Newton did not use the modern integral sign. What was his notation? Did in fact Leibniz choose the elongated "S" symbol to indicate that it is a "smear sum" of infinitesimals?

8.70   Using sectors of infinitesimal angle width $d\theta$, argue that the area of a region in the plane given by a polar equation $r = f(\theta)$, $\alpha \le \theta \le \beta$, is

$$A = \frac{1}{2}\int_{\alpha}^{\beta} f(\theta)^2 \, d\theta.$$

Also, formulate hypotheses and provide a formal Darboux sum proof of this same result.
    Hint: See exercise 5.19.

8.71   Find the area of one leaf of $r = \sin 4\theta$.
    Answer: $A = \pi/16$.

8.72   The *E*-field voltage gain pattern of a certain Yagi antenna is given by
         $r = \cos 2\theta$, $-\pi/4 \leq \theta \leq \pi/4$. Its *half-power beamwidth* is $\theta = \pm\pi/8$
         since power r^2 has dropped off by half by this angle. What percentage
         of all the power transmitted is contained within this half-power
         beamwidth?
              Answer: $1/2 + 1/\pi \approx 82\%$.

8.73   Show that the period p of the planar pendulum $\ddot{\theta} = -(g/a)\sin\theta$ of
         Exercise 6.58 is given by the rule

         $$p = \sqrt{\frac{a}{g}} \int_0^{\theta_0} \frac{d\theta}{\sqrt{\cos\theta - \cos\theta_0}},$$

         where θ_0 is the angle of maximal deflection.
              Outline: Time is the accumulation of infinitesimal displacement
         over rate:

         $$\frac{p}{4} = \int_0^{\theta_0} \frac{ds}{v} = \int_0^{\theta_0} \frac{d\theta}{d\theta/dt}.$$

         Using conservation of energy, rewrite $d\theta/dt$ as a function of θ.

8.74   Investigate the convergence of the improper integral

         $$I = \int_0^\infty \cos x^2 \, dx.$$

8.75   (H. C. Urey) Why are the first several pages of a table of logarithms
         more worn than the last several?

8.76   Prove that if the experimental outcome X is normally distributed
         with mean μ and standard deviation σ, then the outcomes cluster
         about the mean as follows:

         $$\text{prob}(-\sigma < X - \mu < \sigma) \quad \approx 0.6827,$$
         $$\text{prob}(-2\sigma < X - \mu < 2\sigma) \approx 0.9545,$$
         $$\text{prob}(-3\sigma < X - \mu < 3\sigma) \approx 0.9973,$$
         $$\text{prob}(-4\sigma < X - \mu < 4\sigma) \approx 0.9999.$$

         Outline: Reduce to the standard case $\mu = 0$ and $\sigma = 1$ by means of
         the change of variable $Y = (X - \mu)/\sigma$, then proceed numerically.

8.77   Continuing with example 17, what percentage of the students
         received a score between 600 and 700 on a normally distributed test
         outcome with mean $\mu = 500$ and standard deviation $\sigma = 100$?
              Answer: 13.6%.

8.78   Repeatedly perform sixteen tosses of a coin (either manually or via
        simulation). Assign $X = -1$ to tails and $X = 1$ to heads. Add up the
        outcomes X_i and divide by 4 to obtain a new experimental outcome
        Y; that is,

$$Y = \frac{X_1 + X_2 + \cdots + X_{16}}{4},$$

        where X_i is the outcome of the i-th toss, $i = 1, 2, \ldots, 16$. Construct
        a histogram of the many outcomes of Y. Assemble graphical evidence
        that the experimental outcome Y appears normally distributed with
        $\mu = 0$ and $\sigma = 1$.

9

Infinite Series

In this chapter we extend the notion of equality to infinite sums.

9.1 Zeno's Paradoxes

The Sophist teachers of classical Greece challenged their students to explain apparent paradoxes. For example: *An archer launches an arrow toward a far wall. To reach the wall the arrow must first travel halfway to the wall, then half of the remaining distance, then half again, and so on. The arrow cannot reach the wall. Or: Achilles, the fleetest man in Greece, is challanged by a tortoise to a footrace. The tortoise is given a generous head start. To pass the tortoise, Achilles must first reach the starting point of the tortoise. Once reached, he must then reach the new position of the tortoise, and so on. Achilles will lose the race.*

The evident sticking point for the students of the time was the inability to conceive of executing infinitely many discrete actions in a finite amount of time. Two thousand years were to pass before the foundation was laid by Cauchy to explain away these paradoxes.

9.2 Convergence of Sequences

A *sequence* of real numbers is a function $f : \mathbf{N} \longrightarrow \mathbf{R}$, commonly denoted by $\{a_n\}_{n=1}^{\infty}$ where $a_n = f(n)$. Such a sequence *converges to the limit L* if every open neighborhood of L contains all but a finite number of the terms a_n of the sequence $\{a_n\}_{n=1}^{\infty}$; that is, for every $\epsilon > 0$, there exists $N \in \mathbf{N}$ so that

$$n > N \quad \text{implies} \quad |a_n - L| < \epsilon. \qquad (9.1)$$

When the sequence $\{a_n\}_{n=1}^{\infty}$ converges to L we write

$$\lim_{n \to \infty} a_n = L. \qquad (9.2)$$

Example 1. The sequence $a_n = 1/n$ clearly converges to 0, since for any $\epsilon > 0$, merely choose $N = [1/\epsilon]$, where $[x]$ denotes the greatest integer less than or equal to x.

Example 2. The sequence $a_n = r^n$ also converges to 0 whenever $|r| < 1$. This can be seen by taking $N = [\ln \epsilon / \ln |r|]$ (exercise 9.1).

Example 3. The constant sequence $a_n = c$ clearly converges to c. Any N will work.

Under the right metric, one can think of sequential convergence as a sort of continuity at ∞. See exercise 9.2. So it is without suprise that the following results obtain.

Theorem A. *The limit of a sum is the sum of the limits. The limit of a product is the product of the limits.* That is, suppose the sequences $\{a_n\}_{n=1}^{\infty}$ and $\{b_n\}_{n=1}^{\infty}$ both converge. Then so do their sum and product sequences, and in fact

$$\lim_{n \to \infty} (a_n + b_n) = \lim_{n \to \infty} a_n + \lim_{n \to \infty} b_n, \tag{9.3a}$$

$$\lim_{n \to \infty} (a_n \cdot b_n) = \lim_{n \to \infty} a_n \cdot \lim_{n \to \infty} b_n. \tag{9.3b}$$

Proof 1. By exercise 9.3 the map $n \to (a_n, b_n)$ can be extended to a continuous function on $X = \mathbf{N} \cup \{\infty\}$ into $\mathbf{R} \times \mathbf{R}$ under the metric of exercise 9.2. As established in the proofs of theorem D and E of §4.3, addition and multiplication $(x, y) \mapsto x + y$ and $(x, y) \mapsto x \cdot y$ are continuous. Thus the composition maps $n \mapsto a_n + b_n$ and $n \mapsto a_n \cdot b_n$ are continuous (convergent).

Proof 2. The more traditional proof proceeds as follows: Let $\epsilon > 0$ be given and let $\lim_{n \to \infty} a_n = A$ and $\lim_{n \to \infty} b_n = B$. Find $N_a, N_b \in \mathbf{N}$ so that $n > N_a$ implies $|a_n - A| < \epsilon/2$ and $n > N_b$ implies $|b_n - B| < \epsilon/2$. Thus when $n > N = \max (N_a, N_b)$,

$$|a_n + b_n - (A + B)| \leq |a_n - A| + |b_n - B| < \epsilon/2 + \epsilon/2 = \epsilon,$$

proving (9.3a).

Since *convergent sequences are bounded* (exercise 9.5), find β so that $|b_n| < \beta$ for all $n \in \mathbf{N}$. Let $\eta = \epsilon/(\beta + |A|)$. Find $N_a, N_b \in \mathbf{N}$ so that $n > N_a$ implies $|a_n - A| < \eta$ and $n > N_b$ implies $|b_n - B| < \eta$. Then for $n > N = \max (N_a, N_b)$, by standard trick #1,

$$\begin{aligned} |a_n b_n - AB| &= |a_n b_n - A b_n + A b_n - AB| \\ &\leq |a_n - A| \cdot |b_n| + |A| \cdot |b_n - B| < \eta \beta + |A| \eta = \epsilon, \end{aligned}$$

proving (9.3b).

Theorem B. *The limit of a quotient is the quotient of the limits, provided the limit of the denominator is not zero.* That is, if $\lim_{n\to\infty} a_n = A$ and $\lim_{n\to\infty} b_n = B$, where $b_n, B \neq 0$, then

$$\lim_{n\to\infty} \frac{a_n}{b_n} = \frac{A}{B}. \qquad (9.4)$$

Proof. Exercise 9.6.

Theorem C. *A nondecreasing bounded sequence converges.*

Proof. Suppose $a_1 \leq a_2 \leq a_3 \leq \cdots \leq \alpha$. Then (exercise 9.7)

$$\lim_{n\to\infty} a_n = \sup_{n\in\mathbf{N}} a_n. \qquad (9.5)$$

Theorem D. (sequential compactness) *Every bounded sequence possesses a convergent subsequence.*

Proof. Suppose the sequence $\{a_n\}_{n=1}^{\infty}$ is bounded by (say) α, that is, $-\alpha \leq a_n \leq \alpha$ for all $n \in \mathbf{N}$. There must exist a point a of the compact interval $I = [-\alpha, \alpha]$ with the property that every neighborhood of a contains infinitely many terms of the sequence $\{a_n\}_{n=1}^{\infty}$; if not, I possesses a cover of open sets each containing finitely many terms of the sequence, an impossibility since such a cover possesses a finite subcover. Thus for each $k \in \mathbf{N}$, recursively find subscripts $n_k > n_{k-1}$ with $|a_{n_k} - a| < 1/k$, thereby constructing a subsequence with

$$\lim_{k\to\infty} a_{n_k} = a.$$

9.3 Convergence of Series

The infinite series

$$\sum_{k=1}^{\infty} a_k = a_1 + a_2 + a_3 + \cdots \qquad (9.6)$$

is said to *converge to the value s* when the sequence of *partial sums*

$$s_n = \sum_{k=1}^{n} a_k \qquad (9.7)$$

converges to s. When the series (9.6) converges to s we extend equality by writing

$$\sum_{k=1}^{\infty} a_k = s. \tag{9.8}$$

Example 4. As long as $|x| < 1$, the *geometric series*

$$\sum_{k=0}^{\infty} x^k = \frac{1}{1-x}. \tag{9.9}$$

To see this, note that the partial sums of the geometric series can be explicitly given by (exercise 9.8)

$$s_n = \sum_{k=0}^{n} x^k = \frac{1 - x^{n+1}}{1 - x}. \tag{9.10}$$

Thus

$$\lim_{n \to \infty} s_n = \left(\frac{1}{1-x}\right) \lim_{n \to \infty} (1 - x^{n+1}) = \frac{1}{1-x}. \tag{9.11}$$

The n-th term test. *The terms of a convergent series have limit zero. That is,*

$$\sum_{k=1}^{\infty} a_k \quad \text{convergent implies} \quad \lim_{k \to \infty} a_k = 0. \tag{9.12}$$

Proof. Employing our by now generic notation,

$$\lim_{n \to \infty} a_n = \lim_{k \to \infty} (s_n - s_{n-1}) = s - s = 0.$$

However, the converse is patently false!

Example 5. The *harmonic series*

$$\sum_{k=1}^{\infty} \frac{1}{k} = 1 + \frac{1}{2} + \frac{1}{3} + \cdots \tag{9.13}$$

diverges (does not converge) since its n-th partial sum is the upper Darboux sum (and left-hand rule estimate) of the integral of $1/x$

on $[1, n+1]$; that is,

$$s_n = \sum_{k=1}^{n} \frac{1}{k} > \int_1^{n+1} \frac{dx}{x} = \ln(n+1),$$

thus the partial sums s_n diverge to infinity. This is an instance of the use of the integral test of the next section.

9.4 Convergence Tests for Positive Series

Much of the practice of mathematics consists of estimation. There are important results that follow from the mere convergence or divergence of a series. See example 11 below. Our goal here is to list and prove several of the famous tests that enable one to decide the convergence or divergence of positive-term series without explicitly calculating their partial sums. One observation is clear from theorem C: *A series of nonnegative terms is convergent if and only if its partial sums are bounded.*

The comparison test. Suppose $0 \le a_k \le b_k$ for all $k \in \mathbf{N}$. Then

$$\sum_{k=1}^{\infty} b_k \quad \text{convergent implies} \quad \sum_{k=1}^{\infty} a_k \quad \text{convergent,} \qquad (9.14a)$$

$$\sum_{k=1}^{\infty} a_k \quad \text{divergent implies} \quad \sum_{k=1}^{\infty} b_k \quad \text{divergent.} \qquad (9.14b)$$

Proof. If $\sum_k^{\infty} b_k$ is convergent, its sum serves as an upper bound for the partial sums of $\sum_k^{\infty} a_k$.

Example 6. Since term-by-term

$$\sum_{k=1}^{\infty} \frac{1}{k+\sqrt{k}} \ge \sum_{k=1}^{\infty} \frac{1}{k+k} = \frac{1}{2} \sum_{k=1}^{\infty} \frac{1}{k},$$

the first diverges, since any nonzero multiple of the harmonic series (example 5) diverges.

The limit comparison test. Suppose that $a_k \ge 0$, $b_k > 0$, and that $\lim_{k\to\infty} a_k/b_k = L > 0$. Then the two series

$$\sum_{k=1}^{\infty} a_k \quad \text{and} \quad \sum_{k=1}^{\infty} b_k$$

both simultaneously converge or diverge.

Proof. Taking $0 < \epsilon < L$, eventually $0 < L - \epsilon \leq a_k/b_k \leq L + \epsilon$ for all large $k \geq K$ and hence

$$(L - \epsilon) \sum_{k=K}^{N} b_k \leq \sum_{k=K}^{N} a_k \leq (L + \epsilon) \sum_{k=K}^{N} b_k.$$

Example 7. The series

$$\sum_{k=1}^{\infty} \frac{2k}{3k^3 - k^2 + k + 1}$$

converges because (a), as we will see in the next example, the series

$$\sum_{k=1}^{\infty} \frac{1}{k^2}$$

converges, and (b),

$$\lim_{k \to \infty} \frac{3k^3 - k^2 + k + 1}{2k^3} = \frac{3}{2}.$$

The integral test. Suppose $a_k = f(k)$ where f is a continuous positive decreasing function on $[1, \infty)$. Then the series

$$\sum_{k=1}^{\infty} a_n \qquad (9.15a)$$

and the improper integral

$$\int_1^{\infty} f(x)\, dx \qquad (9.15b)$$

both simultaneously converge or diverge.

Proof. Because f is decreasing and $f(k) = a_k$, the terms a_k of the series serve as the right-hand and left-hand rule estimates

$$\int_k^{k+1} f(x)\, dx \leq a_k \leq \int_{k-1}^{k} f(x)\, dx, \quad k > 1,$$

and hence

$$\int_1^{n+1} f(x)\, dx \leq \sum_{k=1}^{n} a_k \leq a_1 + \int_1^{n} f(x)\, dx. \qquad (9.16)$$

Example 8. The *Riemann zeta function*

$$\zeta(s) = \sum_{k=1}^{\infty} \frac{1}{k^s} \tag{9.17}$$

converges for $s > 1$, since

$$\int_1^{\infty} \frac{dx}{x^s} = -\frac{1}{s-1} \cdot \frac{1}{x^{s-1}}\bigg|_1^{\infty} = \frac{1}{s-1}. \tag{9.18}$$

9.5 Convergence Tests for Signed Series

A series $\sum_k^{\infty} a_k$ is said to be *absolutely convergent* when $\sum_k^{\infty} |a_k|$ converges.

Theorem E. *Absolutely convergent series converge.*

Proof. Suppose $\sum_1^{\infty} a_k$ converges absolutely. Then since the nonnegative series $\sum_1^{\infty} (a_k + |a_k|)$ converges (absolutely) by comparison with $\sum_1^{\infty} 2|a_k|$, we may write the partial sums of our original series as a difference of two convergent sequences:

$$\sum_{k=1}^{n} a_k = \sum_{k=1}^{n} (a_k + |a_k|) - \sum_{k=1}^{n} |a_k|. \tag{9.19}$$

Dirichlet's test. Suppose that a_k is a nonincreasing sequence with limit 0 and that the partial sums $B_n = \sum_1^n b_k$ are bounded. Then the series

$$\sum_{k=1}^{\infty} a_k b_k \tag{9.20}$$

converges.

Proof. The following discrete integration by parts is called *Abel summation:* Set the empty sum $B_0 = 0$. Then

$$\sum_{k=1}^{n} a_k b_k = \sum_{k=1}^{n} a_k (B_k - B_{k-1}) = \sum_{k=1}^{n} a_k B_k - \sum_{k=1}^{n} a_k B_{k-1}$$

$$= \sum_{k=1}^{n} a_k B_k - \sum_{k=1}^{n} a_{k+1} B_k + a_{n+1} B_n$$

$$= \sum_{k=1}^{n} (a_k - a_{k+1}) B_k + a_{n+1} B_n. \tag{9.21}$$

But since the B_n are bounded by say β and the a_k are nonincreasing,

$$\sum_{k=1}^{n} |(a_k - a_{k+1})B_k| = \sum_{k=1}^{n} (a_k - a_{k+1})|B_k|$$

$$\le \beta \sum_{k=1}^{n} (a_k - a_{k+1}) = \beta(a_1 - a_{n+1}),$$

and thus the series

$$\sum_{k=1}^{\infty} (a_k - a_{k+1})B_k$$

is absolutely convergent.

We have by Abel's trick broken the partial sum of the series (9.20) into the two terms of (9.21): a partial sum of an absolutely convergent series plus a sequence with limit 0.

Example 9. The *alternating harmonic series*

$$\sum_{k=1}^{\infty} \frac{(-1)^{k+1}}{k} = 1 - \frac{1}{2} + \frac{1}{3} - \frac{1}{4} + \cdots \qquad (9.22)$$

converges since $a_k = 1/k$ is decreasing to 0 while

$$B_n = \sum_{k=1}^{n} (-1)^{k+1} = 1 - 1 + 1 - 1 + \cdots + (-1)^{n+1}$$

takes on only the two values 0 and 1.

The alternating harmonic series is an example of a *conditionally convergent series,* a series that converges but not absolutely. As we shall see, one must be wary of conditionally convergent series. The argument of example 9 generalizes.

The alternating series test. Suppose a_k is a nonincreasing sequence converging to 0. Then the alternating series

$$\sum_{k=1}^{\infty} (-1)^{k+1} a_k = a_1 - a_2 + a_3 - a_4 + \cdots \qquad (9.23)$$

converges.

Example 10. The series

$$\sum_{k=2}^{\infty} \frac{(-1)^k}{\ln k} = \frac{1}{\ln 2} - \frac{1}{\ln 3} + \frac{1}{\ln 4} - \cdots$$

converges (albeit *very* slowly). It does not converge absolutely by the comparison $1/\ln k > 1/k$.

9.6 Manipulating Series

Theorem F. *The result of adding two convergent series term-by-term is a convergent series. Multiplying each term of a convergent series by the same constant results in a convergent series.* In fact, suppose both series $\sum_{1}^{\infty} a_k$ and $\sum_{1}^{\infty} b_k$ converge. Then

$$\sum_{k=1}^{\infty} (a_k + b_k) = \sum_{k=1}^{\infty} a_k + \sum_{k=1}^{\infty} b_k \qquad (9.24a)$$

and

$$\sum_{k=1}^{\infty} c\, a_k = c \sum_{k=1}^{\infty} a_k. \qquad (9.24b)$$

Proof. The result follows from theorem A.

Theorem G. *An absolutely convergent series can be reordered without affecting its convergence or sum.*

Proof. Suppose that $s_\infty = \sum_{1}^{\infty} a_k$ is absolutely convergent. Consider any *reordering* of the terms of this series to form the new series

$$\sum_{k=1}^{\infty} a_{f(k)}$$

given by some bijective $f : \mathrm{N} \to \mathrm{N}$. If $M = \max\{f(1), f(2), \ldots, f(m)\}$,

$$\sum_{k=1}^{m} |a_{f(k)}| \le \sum_{k=1}^{M} |a_k| \le \sum_{k=1}^{\infty} |a_k|,$$

and so the reordered series is also absolutely convergent. Let $\epsilon > 0$ and find n so that

$$\left| s_\infty - \sum_{k=1}^{n} a_k \right| + \sum_{n+1}^{\infty} |a_k| < \epsilon.$$

Let $P = \max\{f^{-1}(1), f^{-1}(2), \ldots, f^{-1}(n)\}$. Then if $p > P$,

$$
\begin{aligned}
\left| s_\infty - \sum_{k=1}^{p} a_{f(k)} \right| &\leq \left| s_\infty - \sum_{k=1}^{n} a_k \right| + \left| \sum_{k=1}^{n} a_k - \sum_{k=1}^{p} a_{f(k)} \right| \\
&\leq \left| s_\infty - \sum_{k=1}^{n} a_k \right| + \left| \sum_{\text{some } k>n} a_k \right| \\
&\leq \left| s_\infty - \sum_{k=1}^{n} a_k \right| + \sum_{\text{some } k>n} |a_k| \\
&\leq \left| s_\infty - \sum_{k=1}^{n} a_k \right| + \sum_{k=n+1}^{\infty} |a_k| < \epsilon.
\end{aligned}
$$

Thus

$$
\sum_{k=1}^{\infty} a_{f(k)} = s_\infty.
$$

The following is a contender for the most bizarre result in all of mathematics. It is an object lesson in why one must always proceed with care.

Theorem H. *A conditionally convergent series can be reordered to add to any prescribed value whatsoever.*

Proof. Let α be any (say) positive number and suppose

$$
s_\infty = \sum_{k=1}^{\infty} a_k \tag{9.25}
$$

is convergent but not absolutely convergent. The two series of positive terms and negative terms of (9.25) must both diverge, that is (exercise 9.15),

$$
\sum_{p=1}^{\infty} a_{k_p} = \infty, \qquad a_{k_p} > 0, \tag{9.26a}
$$

and

$$
\sum_{n=1}^{\infty} a_{k_n} = -\infty, \qquad a_{k_n} < 0. \tag{9.26b}
$$

Reorder (9.25) as follows: Add the first several positive terms a_{k_p} until the sum exceeds α. Then add in the first several negative terms

a_{k_n} until the combined sum is less than α. Now add the next several positive terms to exceed α, and so on. All terms of the original sum (9.25) will be used, yet the sum in the limit is α.

Theorem I. *The terms of a convergent series can be grouped arbitrarily without affecting its convergence or sum.*

Proof. Suppose that $s_\infty = \sum_1^\infty a_k$ converges and that $k_1 < k_2 < k_3 < \cdots$ is an increasing sequence of natural numbers. Let

$$c_n = \sum_{i=k_{n-1}+1}^{k_n} a_i. \tag{9.27}$$

Then the series of grouped terms

$$\sum_{n=1}^\infty c_n = (a_1 + \cdots + a_{k_1}) + (a_{k_1+1} + \cdots + a_{k_2}) + \cdots \tag{9.28}$$

also converges to s_∞, since its partial sums form a subsequence of the partial sums of the original series—see exercise 9.16.

Theorem J. *The result of multiplying out two absolutely convergent series is again an absolutely convergent series that converges to the product of the sums of the two factors.*

Proof. Suppose that $s_\infty = \sum_1^\infty a_k$ and $t_\infty = \sum_1^\infty b_k$ converge absolutely. Consider any bijection $f : \mathbf{N} \times \mathbf{N} \longrightarrow \mathbf{N}$ and the resulting ordering of the product series

$$\left(\sum_{i=1}^\infty a_i\right)\left(\sum_{j=1}^\infty b_j\right) = \sum_{\substack{p=1 \\ p=f(i,j)}}^\infty a_i b_j = \sum_{p=1}^\infty c_p, \tag{9.29}$$

where $c_p = a_i b_j$ when $p = f(i, j)$.

For each $P \in \mathbf{N}$, let N be the maximum of either entry of the list of P 2-tuples $f^{-1}(\{1, 2, \ldots, P\})$. Then

$$\sum_{p=1}^P |c_p| = \sum_{\substack{p=1 \\ p=f(i,j)}}^P |a_i b_j| \leq \left(\sum_{i=1}^N |a_i|\right)\left(\sum_{j=1}^N |b_j|\right)$$

and thus the product series (9.29) converges absolutely and can be reordered by Theorem G so that its convergence and value are independent of the numbering scheme f.

Since

$$S_n = \left(\sum_{i=1}^{n} a_i \right) \left(\sum_{j=1}^{n} b_j \right) = \sum_{1 \le i,j \le n} a_i b_j$$

(under some systematic ordering scheme) is a subsequence of partial sums of (some ordering of) the product series $\sum_p c_p$, we see that

$$\lim_{n \to \infty} \left(\sum_{i=1}^{n} a_i \right) \left(\sum_{j=1}^{n} b_j \right) = \sum_{p=1}^{\infty} c_p = s_\infty t_\infty. \tag{9.30}$$

Example 11. Suppose there are only a finite number of primes $p \in \mathbf{N}$. By theorem J, when $s > 1$, we may multiply the (absolutely convergent) geometric series

$$\frac{1}{1 - 1/p^s} = \sum_{k=0}^{\infty} \frac{1}{p^{ks}}$$

to obtain, because of unique factorization, the famous identity that is the principal avenue of attack on the prime numbers:

$$\left(1 - \frac{1}{2^s} \right)^{-1} \left(1 - \frac{1}{3^s} \right)^{-1} \left(1 - \frac{1}{5^s} \right)^{-1} \cdots = \sum_{k=1}^{\infty} \frac{1}{k^s}. \tag{9.31}$$

By assumption, the product on the left of (9.31) is a finite product and thus has a finite value at $s = 1$. But the right-hand side, the Riemann zeta function of example 8, has an infinite limit as $s \to 1^+$ (exercise 9.18). Thus there are infinitely many primes.

We could reach the same conclusion by proving that $\zeta(2) = \pi^2/6$ is irrational (exercise 9.47) while the left-hand side of (9.31)—if a finite product—is clearly rational at $s = 2$.

9.7 Power Series

A *power series*

$$f(x) = \sum_{k=0}^{\infty} c_k (x - a)^k \tag{9.32}$$

will define a function f at each point x where the series is convergent.

Theorem K. If the series (9.32) converges at $x = x_0 \ne a$, then it converges absolutely at each point x in the open ball given by $|x - a| < |x_0 - a|$.

Proof. Without loss of generality we may assume $a = 0$. Since the series converges at x_0, the terms $c_k x_0^k \to 0$ and hence are bounded, say $|c_k x_0^k| \le \beta$. Thus, when $|x| < |x_0|$, we have this comparison:

$$\sum_{k=0}^{\infty} |c_k x^k| = \sum_{k=0}^{\infty} |c_k x_0^k| \cdot \left| \frac{x}{x_0} \right|^k \le \beta \sum_{k=0}^{\infty} \left| \frac{x}{x_0} \right|^k = \frac{\beta}{1 - |x/x_0|}.$$

Corollary. The power series (9.32) possesses a *radius R of convergence*. That is, for each power series (9.32) there exists a (unique) R so that the series converges everywhere on the open ball $\{|x - a| < R\}$ and nowhere on its exterior $\{|x - a| > R\}$.

Example 12. The geometric series

$$f(x) = \sum_{k=0}^{\infty} x^k \tag{9.33}$$

has radius of convergence $R = 1$, since it converges at each $|x| < 1$ to $1/(1 - x)$ and since its terms x^k do not have limit 0 when $|x| \ge 1$.

Example 13. The power series

$$f(x) = \sum_{k=1}^{\infty} \frac{x^k}{k} \tag{9.34}$$

can have radius of convergence R at most 1, since the series diverges at $x = 1$. On the other hand, when $x = -1$, the series becomes the (convergent) alternating harmonic series and so (9.34) converges for all $|x| < 1$. Thus $R = 1$.

Note that the *interval of convergence* of the series (9.34) is the half-open interval $[-1, 1)$, while in contrast, the interval of convergence of (9.33) is the open interval $(-1, 1)$.

Example 14. The power series

$$f(x) = \sum_{k=1}^{\infty} k^k x^k \tag{9.35}$$

fails the n-term test when $x \ne 0$. Thus its radius of convergence $R = 0$.

Example 15. The power series

$$f(x) = \sum_{k=1}^{\infty} \frac{x^k}{k^k} \tag{9.36}$$

converges everywhere ($R = \infty$), since eventually its terms $|x/k| < 1$ and thus is convergent by comparison with the geometric series.

Theorem L. *A power series represents a continuous function on the interior of its interval of convergence.*

Proof. Suppose

$$f(x) = \sum_{k=0}^{\infty} c_k x^k$$

has radius of convergence R. Then for $|x_0|, |x| < r < R$,

$$|f(x) - f(x_0)| \leq \left| \sum_{1}^{n} c_k (x - x_0)^k \right| + \left| \sum_{k=n+1}^{\infty} c_k x^k \right| + \left| \sum_{k=n+1}^{\infty} c_k x_0^k \right|$$

$$\leq \left| \sum_{1}^{n} c_k (x - x_0)^k \right| + 2 \sum_{k=n+1}^{\infty} |c_k| r^k.$$

Let $\epsilon > 0$ be given. Because of the absolute convergence of our original series at $x = r$, we may find a large n so that the second term is less than $\epsilon/2$. By polynomial continuity we may find a δ neigborhood of x_0 where the first term is at most $\epsilon/2$.

9.8 Convergence Tests for Power Series

We begin with the most useful of all power series tests.

The ratio test. Suppose eventually that $c_k \neq 0$ and that

$$\lim_{k \to \infty} \left| \frac{c_{k+1}}{c_k} \right| = \rho. \tag{9.37}$$

Then the power series

$$\sum_{k=0}^{\infty} c_k (x - a)^k \tag{9.38}$$

has radius R of convergence

$$R = \frac{1}{\rho}. \tag{9.39}$$

Proof. We may assume $a = 0$. Suppose $\rho \neq 0, \infty$. Since convergence is determined by the tail end of a series, we may also assume that $c_k \neq 0$ and

$$\rho - \epsilon < |c_{k+1}/c_k| \leq \rho + \epsilon \tag{9.40}$$

for all $k \geq 0$ and a fixed $0 < \epsilon < \rho$. But then (exercise 9.19)

$$|c_0|(\rho - \epsilon)^k < |c_k| < |c_0|(\rho + \epsilon)^k. \tag{9.41}$$

Therefore term-by-term,

$$|c_0| \sum_{k=0}^{\infty} |(\rho - \epsilon)x|^k \leq \sum_{k=0}^{\infty} |c_k x^k| \leq |c_0| \sum_{k=0}^{\infty} |(\rho + \epsilon)x|^k. \tag{9.42}$$

Thus our series converges absolutely when $|x| < 1/(\rho + \epsilon)$ and diverges when $|x| > 1/(\rho - \epsilon)$ by comparison with the geometric series on either side of (9.42). But $\epsilon > 0$ is arbitrary. Hence $R = 1/\rho$. The cases $\rho = 0, \infty$ are left as exercise 9.28.

Example 13 (revisited). The power series

$$\sum_{k=1}^{\infty} \frac{x^k}{k}$$

has radius of convergence $R = 1$, since

$$\lim_{k \to \infty} \frac{|c_{k+1}|}{|c_k|} = \lim_{k \to \infty} \frac{1/(k+1)}{1/k} = \lim_{k \to \infty} \frac{k}{k+1} = 1.$$

Example 17. The power series

$$\sum_{k=0}^{\infty} \frac{x^k}{k!} \tag{9.43}$$

converges everywhere ($R = \infty$), since

$$\lim_{k \to \infty} \frac{|c_{k+1}|}{|c_k|} = \lim_{k \to \infty} \frac{k!}{(k+1)!} = \lim_{k \to \infty} \frac{1}{k+1} = 0.$$

The following is satisfying but not often useful.

Hadamard's formula. The series $\sum_0^{\infty} c_k(x - a)^k$ has radius of convergence R determined by

$$\frac{1}{R} = \limsup_k |c_k|^{1/k}. \tag{9.44}$$

Proof. This follows from the ideas of exercises 9.26–9.27.

9.9 Manipulation of Power Series

Two power series may of course be added termwise on their common intervals of convergence. Because power series converge absolutely on the interior of their interval of convergence, two power series may be multiplied, reordered, and grouped to form a convergent power series within their common intervals of convergence:

$$\left(\sum_{k=0}^{\infty} a_k x^k\right) \left(\sum_{k=0}^{\infty} b_k x^k\right) = \sum_{k=0}^{\infty} c_k x^k, \tag{9.45}$$

where the coefficients c_k are given by *cauchy product:*

$$c_k = \sum_{i=1}^{k} a_i b_{k-i} = a_0 b_k + a_1 b_{k-1} + \cdots + a_k b_0. \tag{9.46}$$

Theorem M. *A function given by a power series may be integrated term-by-term within the interior of its interval of convergence.*

Proof. Suppose that

$$f(x) = \sum_{k=0}^{\infty} c_k x^k \tag{9.47}$$

has radius of convergence R and that $-R < -r < a < b < r < R$. Then

$$\int_a^b f(x)\, dx = \int_a^b \sum_{k=0}^{n} c_k x^k\, dx + \int_a^b \sum_{k=n+1}^{\infty} c_k x^k\, dx$$

$$= \sum_{k=0}^{n} c_k \frac{b^{k+1} - a^{k+1}}{k+1} + \int_a^b \sum_{k=n+1}^{\infty} c_k x^k\, dx. \tag{9.48}$$

But since $|c_k r^k| \leq \beta$ for some β, we have for $x \in [a, b]$ that

$$\left| \sum_{k=n+1}^{\infty} c_k x^k \right| \leq \sum_{k=n+1}^{\infty} |c_k r^k| \cdot |x/r|^k$$

$$\leq \beta \sum_{k=n+1}^{\infty} |x/r|^k \leq \beta \left|\frac{x}{r}\right|^{n+1} \frac{r}{|r-x|}.$$

Thus the integral on the right of (9.48) can be made a small as desired for large n. Thus we may pass the integral sign across the infinite sum.

Example 18. For $|x| < 1$,

$$\text{Arctan } x = \int_0^x \frac{dt}{1+t^2} = \int_0^x \sum_{k=0}^{\infty} (-1)^k t^{2k} \, dt$$

$$= \sum_{k=0}^{\infty} \frac{(-1)^k x^{2k+1}}{2k+1} = 1 - \frac{x^3}{3} + \frac{x^5}{5} - \frac{x^7}{7} + \cdots .$$

Theorem N. *A function given by a power series may be differentiated term-by-term on the interior of its interval of convergence.*

Proof. Suppose that

$$f(x) = \sum_{k=0}^{\infty} c_k x^k \tag{9.49}$$

has radius of convergence R and let

$$g(x) = \sum_{k=1}^{\infty} k c_k x^{k-1} \tag{9.50}$$

be the series obtained by differentiating (9.49) term-by-term. The differentiated series (9.50) also has radius of convergence R (exercise 9.42). But then by theorem M, for $|x| < R$,

$$\int_0^x g(t) \, dt = \int_0^x \sum_{k=1}^{\infty} k c_k t^{k-1} \, dt = \sum_{k=1}^{\infty} c_k x^k = f(x) - c_0.$$

Thus $f'(x) = g(x)$.

Example 19. For $|x| < 1$,

$$D \sum_{k=1}^{\infty} \frac{x^k}{k} = \sum_{k=1}^{\infty} x^{k-1} = \sum_{k=0}^{\infty} x^k = \frac{1}{1-x}.$$

Thus (exercise 9.43)

$$\sum_{k=1}^{\infty} \frac{x^k}{k} = -\ln(1-x). \tag{9.51}$$

Consequently by *Abel's theorem* (exercise 9.44)

$$1 - \frac{1}{2} + \frac{1}{3} - \frac{1}{4} + \cdots = \ln 2. \tag{9.52}$$

Corollary. If a function f is given within $|x - a| < R$ by a convergent power series

$$f(x) = \sum_{k=0}^{\infty} c_k(x - a)^k,$$

then its coeficients are uniquely determined by the formula

$$c_n = \frac{f^{(n)}(a)}{n!}. \tag{9.53}$$

Proof. Differentiate the series n times and specialize at $x = a$.

9.10 Taylor Series

A function f that can be given as a convergent power series

$$f(x) = \sum_{k=0}^{\infty} c_k(x - a)^k$$

on an open neighborhood of $x = a$ is said to be *analytic* at a. Which of the familiar functions of calculus are analytic and where? The answer is given by a workhorse of mathematics.

Theorem O. (Taylor's formula) Suppose f has $n+1$ derivatives on some open neighborhood U of $x = a$. Then for each $x \in U$, there is a value c between a and x for which

$$f(x) = \sum_{k=0}^{n} f^{(k)}(a)\frac{(x - a)^k}{k!} + E_n(x, a), \tag{9.54}$$

where

$$E_n(x, a) = f^{(n+1)}(c)\frac{(x - a)^{n+1}}{(n + 1)!}. \tag{9.55}$$

Alternatively, when $f^{(n+1)}$ is continuous, the n-th error term E_n can also be expressed in integral form:

$$E_n(x, a) = \frac{1}{n!} \int_a^x f^{(n+1)}(t)(x - t)^n \, dt. \tag{9.56}$$

Proof. See the proof scheme of exercise 5.20. When $f^{(n+1)}$ is continuous, there is a second trick proof: By repeated integration by parts,

$$
f(b) = f(a) + \int_a^b f'(x)(b-x)^0 \, dx
$$

$$
= f(a) - f'(x)(b-x)\big|_a^b + \int_a^b f''(x)(b-x) \, dx
$$

$$
= f(a) + f'(a)(b-a) - \frac{f''(x)(b-x)^2}{2!}\bigg|_a^b + \frac{1}{2!}\int_a^b f'''(x)(b-x)^2 \, dx = \cdots
$$

$$
= \sum_{k=0}^n f^{(k)}(a)\frac{(b-a)^k}{k!} + \frac{1}{n!}\int_a^b f^{(n+1)}(x)(b-x)^n \, dx. \tag{9.57}
$$

But since $f^{(n+1)}$ is continuous, by exercise 9.46 there is some c between a and b such that

$$
\int_a^b f^{(n+1)}(x)(b-x)^n \, dx = f^{(n+1)}(c)\int_a^b (b-x)^n \, dx,
$$

giving the error formula (9.55).

Example 20. Since the trigonometric functions $\cos x$ and $\sin x$ have derivatives bounded by 1, the error (9.55) becomes for each x,

$$
E_n(x,a) = f^{(n+1)}(c)\frac{(x-a)^{n+1}}{(n+1)!} << \frac{|x-a|^{n+1}}{(n+1)!} \longrightarrow 0
$$

as $n \to \infty$. Thus in particular, expanding about $a = 0$ yields that for all x,

$$
\cos x = 1 - \frac{x^3}{3!} + \frac{x^5}{5!} - \cdots \tag{9.58a}
$$

and

$$
\sin x = x - \frac{x^2}{2!} + \frac{x^4}{4!} - \cdots . \tag{9.58b}
$$

Example 21. We have the everywhere valid expansion

$$e^x = 1 + x + \frac{x^2}{2!} + \frac{x^3}{3!} + \cdots . \tag{9.59}$$

To see that this is correct, note that the error term (9.55) in this case becomes for $a = 0$ and a fixed x,

$$E_n = e^c \frac{x^{n+1}}{(n+1)!} << 3^{|x|} \frac{|x|^{n+1}}{(n+1)!} \longrightarrow 0$$

as $n \to \infty$, validating the expansion (9.59).

As a further consequence, for all x,

$$\cosh x = \frac{e^x + e^{-x}}{2} = 1 + \frac{x^3}{3!} + \frac{x^5}{5!} + \cdots \tag{9.60a}$$

and

$$\sinh x = \frac{e^x - e^{-x}}{2} = x + \frac{x^2}{2!} + \frac{x^4}{4!} + \cdots . \tag{9.60b}$$

Example 22. It is not always easy to obtain a Taylor series expansion in the obvious way via (9.54). For instance, when $f(x) = \sin^2 x$, trickery is preferred: Use instead that $\sin^2 x = (1 - \cos 2x)/2$. Or, the first several terms of the Taylor series for $f(x) = \tan x$ can be obtained by long division:

$$\tan x = \frac{\sin x}{\cos x} = \frac{x - x^3/3! + x^5/5! - \cdots}{1 - x^2/2! + x^4/4! - \cdots}$$
$$= x + \frac{1}{3}x^3 + \frac{2}{15}x^5 + \cdots . \tag{9.61}$$

Warning. An infinitely differentiable function may not be analytic!

Example 23. Consider

$$f(x) = \begin{cases} e^{-1/x^2} & \text{if } x \neq 0 \\ 0 & \text{if } x = 0. \end{cases} \tag{9.62}$$

Then (exercise 9.52) $0 = f(0) = f'(0) = f''(0) = f'''(0) = \cdots$. Hence f has the zero series as Taylor series at $x = 0$. The function f equals its error term (9.55) rather than its Taylor series.

Exercises

9.1   Prove that $\lim_{n\to\infty} r^n = 0$ when $|r| < 1$.

9.2   Consider the set $X = \mathbf{N} \cup \{\infty\}$. We revisualize the points n of X as
        points p_n on the unit circle, where p_n is the intersection of the line
        from the north pole N (now thought of as ∞) to the original point n.

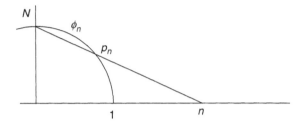

So each natural number $n \mapsto \phi_n$ where ϕ_n is the arclegth from the
zenith down to the point of intersection p_n and $\infty \mapsto 0$. Prove that

$$\frac{1 + \cos \phi_n}{\sin \phi_n} = n.$$

Check that X becomes a (compact) metric space under the distance
function given by angular distance, that is, $\mathrm{d}(m, n) = |\phi_m - \phi_n|$
and $\mathrm{d}(n, \infty) = \phi_n$.

9.3   Let $X = \mathbf{N} \cup \{\infty\}$. Consider any sequence $f = \{a_n\}_{n=1}^{\infty}$. Extend f to all
        of X by setting $f(\infty) = a_\infty = L$. Prove that under the metric of
        exercise 9.2, $f = \{a_n\}_{n=1}^{\infty}$ converges to L if and only if $f : X \longrightarrow \mathbf{R}$ is
        (everywhere) continuous.

9.4   Verify the details in proof 1 of theorem A.

9.5   Prove that convergent sequences are bounded.
            Hint: All but a finite number of the terms are within 1 of the limit.
            (Alternatively, the continuous image of a compact set is compact
            and hence bounded.)

9.6   Prove theorem B.

9.7   Verify (9.5).

9.8   Verify the formula (9.10) for finite geometric sums.

9.9   **(Project)** Report on the contributions of Baron Augustin-Louis Cauchy to the foundation of sequences and series.

9.10   Compute

$$\sum_{k=1}^{\infty} \left(\frac{-2}{3} \right)^k.$$

9.11   Find the fraction with decimal expansion $0.937937937\cdots$ by writing the decimal expansion as a geometric series.

9.12   Construct an elementary proof of the alternating series test.
        Suggestion: Note that $s_1 \geq s_3 \geq \cdots$ while $s_2 \leq s_4 \leq \cdots$ and $s_{2k-1} \geq s_{2k}$. Apply theorem C.

9.13   Show that

$$\lim_{n\to\infty} \left(1 + \frac{x}{n}\right)^n = e^x.$$

Outline: Find the limit of the logarithms of the terms instead. Write the logarithm as an integral so that the n-th term becomes the product of x with the average value of $1/t$ on $[1, 1 + x/n]$. Or use L'Hôpital's Rule (exercise 9.38).

9.14   Prove *Abel's test:* Suppose that a_k is a *monotone* (nondecreasing or nonincreasing) convergent sequence and that $\sum_1^\infty b_k$ is convergent. Then $\sum_1^\infty a_k b_k$ converges.
        Hint: Track through the proof of Dirichlet's test.

9.15   Show that the series of the positive (negative) terms of a conditionally convergent series must diverge.

9.16   Prove that each subsequence of a convergent sequence converges to the same identical limit.

9.17   Using a calculating machine, compute the first thousand partial sums of the harmonic series. Why does the series appear to converge?

9.18   Show that for the Riemann zeta function (example 8),

$$\lim_{s\to 1^+} \zeta(s) = \infty.$$

9.19   Verify (9.41).

9.20   (Sequential continuity) Prove that f is continuous at $x = a$ if and only if for every sequence $\{a_n\}_{n=1}^{\infty}$ of values drawn from the domain of f that converges to a we have

$$\lim_{n \to \infty} f(a_n) = f(a).$$

9.21   Establish the convergence of the improper integral

$$I = \int_0^{\infty} \frac{\sin x}{x} \, dx.$$

Outline: Divide the integral into a sum of integrals over the intervals $[(n-1)\pi, n\pi]$, where $n \in \mathbf{N}$. Apply the alternating series test.

9.22   Design a scheme for stacking bricks, one per level, to make a one-sided arch that extends beyond any prescribed overhang. (The arch will collapse unless, at each level, the center of mass of the bricks above lies above the brick at that level.)

   Hint: Use the terms of the harmonic series written backwards as the successive overhang distances.

9.23   Explain away one of Zeno's paradoxes of §9.1.

9.24   Sum the complex series $1 + i/2 - 1/4 - i/8 + 1/16 + i/32 - 1/64 - i/128 + 1/256 - \cdots$ in two ways: by summing up the real and imaginary parts and by using the geometric series formula (9.9).

9.25   Construct an elementary proof that there are infinitely many prime numbers.

   Outline: Suppose $p_1, p_2, \ldots, p_n$ is a complete list. Factor $N = p_1 p_2 \cdots p_n + 1$.

9.26   Prove the *root test:* If eventually $|a_k|^{1/k} < \rho < 1$, then the series $\sum_1^{\infty} a_k$ converges absolutely.

9.27   Strengthen exercise 9.26 by proving that if $\limsup_k |a_k|^{1/k} < 1$, then the series $\sum_1^{\infty} a_k$ converges absolutely.

   Explanation: The *limit superior* of a sequence is its eventual least upper bound, that is, the least of the least upper bounds when finitely many terms are discarded, that is, the largest accumulation point. Formally,

$$\limsup_k b_k = \inf_k \sup_{p \geq k} b_p.$$

9.28   Finish the proof of the ratio test by handling the cases $\rho = 0, \infty$.

9.29   Find the radius of convergence of the *n-th Bessel function of the first kind*

$$J_n(x) = \sum_{k=0}^{\infty} \frac{(-1)^k}{k!\,(n+k)!} \left(\frac{x}{2}\right)^{2k+n}.$$

9.30   The sequence $\{a_k\}_{k=1}^{\infty}$ is *Cesaro summable* if the limit of its running averages

$$s_n = \frac{a_1 + a_2 + \cdots + a_n}{n}$$

exists. Prove that every convergent sequence is Cesaro summable. Prove by example that the converse is false. Thus Cesaro summability is a weakened notion of sequential convergence.

9.31   A sequence $\{a_k\}_{k=1}^{\infty}$ is *cauchy convergent* if all its terms are eventually close, that is, for every $\epsilon > 0$ there exists a natural number N such that $p, q > N$ implies $|a_p - a_q| < \epsilon$. Prove that *a convergent sequence is cauchy convergent.*

9.32   Prove that **R** is complete, that is, *every cauchy convergent sequence converges.*
        Outline: A cauchy convergent sequence is bounded and hence possesses a convergent subsequence whose limit is the limit of the original sequence.

9.33*   Assume the first three axioms of §2.1 for **R**, that cauchy sequences converge, and that the sequence $\{2^{-n}\}_n^{\infty}$ converges to 0. Deduce the supremum axiom IV: *Every nonempty subset S of* **R** *that is bounded from above possesses a supremum.*
        Hint: Choose $a \in S$ and b an upper bound of S. Perform repeated bisections starting with $[a, b]$.

9.34   Expand $f(x) = e^{x^2}$ into its Taylor series about $a = 0$.

9.35   Expand $f(x) = \text{Arctan } x$ into its Taylor series about $a = 0$. Why does the expansion fail to represent f when $|x| > 1$?

9.36   Prove *L'Hôpital's easy rule:* Suppose that $f(x_0) = g(x_0) = 0$, that both $f'(x_0)$ and $g'(x_0)$ exist, and that $g'(x_0) \neq 0$. Then

$$\lim_{x \to x_0} \frac{f(x)}{g(x)} = \frac{f'(x_0)}{g'(x_0)}.$$

For example,

$$\lim_{x \to 0} \frac{\sin x}{x} = \frac{\cos x}{1} \bigg|_{x=0} = 1.$$

Outline:

$$\frac{f'(x_0)}{g'(x_0)} = \frac{\lim_{x \to x_0} (f(x) - f(x_0))/(x - x_0)}{\lim_{x \to x_0} (g(x) - g(x_0))/(x - x_0)}.$$

9.37 Prove *L'Hôpital's rule:* Suppose that both f and g are differentiable on an open neighborhood of $x = x_0$, that both vanish at $x = x_0$, and that g and g' are nonzero on a punctured neighborhood of x_0. Then

$$\lim_{x \to x_0} \frac{f(x)}{g(x)} = \lim_{x \to x_0} \frac{f'(x)}{g'(x)},$$

provided the right limit exists.

Outline: Apply the mean value theorem to $h(x) = f(x)g(x_1) - f(x_1)g(x)$ to obtain that

$$\frac{f(x_1)}{g(x_1)} = \frac{f'(\xi)}{g'(\xi)}$$

for some ξ between x_1 and x_0. Let x_1 approach x_0.

9.38 Prove L'Hôpital's rule at ∞: Suppose f and g are differentiable on $I = (a, \infty)$, that both have either limit 0 or ∞ at ∞, and that g, g' are nonzero on I. Then

$$\lim_{x \to \infty} \frac{f(x)}{g(x)} = \lim_{x \to \infty} \frac{f'(x)}{g'(x)},$$

provided the right limit exists.

9.39 Use L'Hôpital's rule to verify the following limits:

a. $\displaystyle \lim_{x \to \infty} \frac{3x^2 - x + 1}{x^2 + 5x + 2} = 3.$

b. $\displaystyle \lim_{n \to \infty} \left(1 - \frac{x}{n}\right)^n = e^{-x}.$

c. For $\alpha > 0$, $\displaystyle \lim_{x \to 0} x^\alpha \ln x = 0.$

9.40 Determine the convergence or divergence:

a. $\displaystyle \sum_{k=1}^{\infty} \frac{2^k}{k^2}$

b. $\displaystyle \sum_{k=1}^{\infty} \left(1 - \frac{1}{k}\right)^k$

c. $\displaystyle\sum_{k=1}^{\infty} \frac{(-1)^k \ln k}{k + \ln k}$

d. $\displaystyle\sum_{k=1}^{\infty} k^2 e^{-k}$

e. $\displaystyle\sum_{k=1}^{\infty} \frac{k!}{k^k}$

f. $\displaystyle\sum_{k=2}^{\infty} \frac{k}{\ln^k k}$

g. $\displaystyle\sum_{k=1}^{\infty} \frac{\cos k\pi}{\sqrt{k}}$

h. $\displaystyle\sum_{k=2}^{\infty} \frac{1}{k \ln k}$.

9.41 Find the interval of convergence:

a. $\displaystyle\sum_{k=0}^{\infty} \frac{2^k (x-1)^k}{k}$

b. $\displaystyle\sum_{k=0}^{\infty} \sqrt{k}(2x-1)^k$

c. $\displaystyle\sum_{k=0}^{\infty} \frac{x^k}{k \ln k}$

d. $\displaystyle\sum_{k=0}^{\infty} \ln^k x$.

9.42 Show that differentiating a power series term-by-term does not change its radius of convergence $R > 0$.
   Outline: For $|x| < r < R$,

$$\sum_{k=1}^{\infty} |kc_k x^{k-1}| \le \sum_{k=1}^{\infty} |c_k r^{k-1}| \cdot k|x/r|^{k-1} \le \frac{\beta}{r} \sum_{k=1}^{\infty} k|x/r|^{k-1}.$$

   Apply the ratio test.

9.43 Verify (9.51).

9.44*   Prove *Abel's theorem:* Suppose $\sum_k^\infty c_k x^k$ converges at $x = r > 0$. Then

$$\lim_{x \to r^-} \sum_{k=0}^\infty c_k x^k = \sum_{k=0}^\infty c_k r^k.$$

Deduce (9.52).

Outline: We may assume $r = 1$. Let $C_n = \sum_n^\infty c_k$. Then for $0 < x < 1$, by Abel summation

$$\sum_{k=n}^\infty c_k(1 - x^k) = \sum_{k=n}^\infty C_k(x^k - x^{k-1}) + C_n(1 - x^{n-1}) << 2 \sup_{k \ge n} |C_k|.$$

Thus the tail end of $\sum_0^\infty c_k(1 - k^k)$ can be made arbitrarily small independently of x for large n.

9.45   Show

$$1 - \frac{1}{3} + \frac{1}{5} - \frac{1}{7} + \cdots = \frac{\pi}{4}.$$

Hint: Integrate the geometric series expansion of $1/(1 + x^2)$.

9.46   Prove that if f is continuous and g is integrable of constant sign on $[a, b]$, then for some $a < c < b$,

$$\int_a^b f(x)g(x)\, dx = f(c) \int_a^b g(x)\, dx.$$

9.47*   Prove that

$$1 + \frac{1}{2^2} + \frac{1}{3^2} + \frac{1}{4^2} + \cdots = \frac{\pi^2}{6}.$$

9.48*   Prove the *contraction mapping theorem:* Suppose X is a *complete* metric space under the metric $d(\cdot, \cdot)$, that is, every cauchy sequence drawn from X converges to a point of X. Suppose $f : X \to X$ a *contraction mapping* of X, that is, for some positive $\alpha < 1$, $d(f(x), f(y)) \le \alpha\, d(x, y)$ for all $x, y \in X$. Then f posseses exactly one *fixed point* x_∞, that is, a point with $f(x_\infty) = x_\infty$.

Outline: Start with any $x_0 \in X$ and interate $x_{n+1} = f(x_n)$. Then $d(x_{n+1}, x_n) \le \alpha d(x_n, x_{n-1}) \le \cdots \le \alpha^n d(x_1, x_0)$. Thus

$$d(x_n, x_m) \le d(x_1, x_0) \sum_{k=m}^{n-1} \alpha^k,$$

giving that $\{x_n\}_{n=1}^\infty$ is a cauchy sequence with limit x_∞.

9.49   Suppose the differentiable $f : [a, b] \to [a, b]$ has $|f'(x)| \le \alpha < 1$ on
        $[a, b]$. Show f has exactly one fixed point.
            Example: $f(x) = \cos x$ on $[0, 1]$. See exercise 4.27.

9.50   Suppose that $f : \mathbf{R} \longrightarrow \mathbf{R}$ is continuously differentiable on an open
        neighborhood of a fixed point x_∞ and that $|f'(x_\infty)| < 1$. Prove that
        about x_∞ is a basin of attraction U where iterations x, $f(x)$, $f(f(x))$,
        $f(f(f(x)))$, ... for any choice of $x \in U$ will lead to the fixed point x_∞.

9.51   (Cournot) Two companies sell the same commodity. Both follow the
        identical strategy: If the competitor sells q units, then price to sell
        $f(q)$ units. Prove that if $|f'(q)| \le \alpha < 1$ everywhere, then the price
        competition will lead eventually to both companies selling the same
        number of units.

9.52   Establish that the function f of (9.61) has derivatives of all orders
        everywhere, and in particular, $f^{(n)}(0) = 0$ for all $n \ge 0$.

9.53* (Weierstrass) Produce an everywhere continuous function that is
        nowhere differentiable.
            Outline: At each rational number r_n, construct the triangular wave

$$f_n(x) = \begin{cases} 2^{-n} - |x - r_n| & \text{if } |x - r_n| \le 2^{-n} \\ 0 & \text{otherwise.} \end{cases}$$

        Superimpose the f_n to form *van der Waerden's example:*

$$f(x) = \sum_{n=1}^{\infty} f_n(x).$$

9.54   Prove that if the *Dirichlet series*

$$f(s) = \sum_{k=1}^{\infty} \frac{a_k}{k^s}$$

        converges at $s = s_0$, then it converges for $s \ge s_0$ and converges
        absolutely for $s > s_0 + 1$.
            Hint: Apply Dirichlet's test.

9.55   Referring to §7.6, show that the set of rationals is a Borel set of
        measure zero; that is, $\mu(\mathbf{Q}) = 0$.

9.56   Suppose f is a bounded Borel-measurable function on $[a, b]$. In
        contrast to the construction of the Riemann integral, partition $[a, b]$

into finitely many disjoint Borel sets:

$$\mathcal{P} = \bigcup_{i=1}^{n} E_i,$$

and define the upper and lower Darboux sums

$$U(\mathcal{P}) = \sum_{i=1}^{n} \sup_{E_i} f(x)\, \mu(E_i)$$

and

$$L(\mathcal{P}) = \sum_{i=1}^{n} \inf_{E_i} f(x)\, \mu(E_i).$$

As in (7.6), show that no lower sum of one partition can exceed an upper sum of another. If all upper and lower sums trap but one number I, we say that f is *Lebesgue integrable* and write

$$I = \int_a^b f(x)\, d\mu(x).$$

Show that Dirichlet's example (example 1 of §7.1) is Lebesgue integrable.

References

M. Artin, *Algebra*, Prentice Hall, Upper Saddle River, NJ, 1991.

A. M. Bruckner, J. B. Bruckner, and B. S. Thomson, *Real Analysis*, Prentice Hall, Upper Saddle River, NJ, 1997.

G. Cantor, *Contributions to the Founding of the Theory of Transfinite Numbers*, Dover Publications, NY, 1955.

CRC, *CRC Standard Mathematical Tables and Formulae*, 31st edition, ed. D. Zwillinger, 2002.

P. C. W. Davies, *Quantum Mechanics*, Chapman and Hall, NY, 1989.

Euclid, *The Elements*, Dover Publications, NY, 1956.

J. Gleick, *Issac Newton*, Vintage Books (Random House), NY, 2003.

P. R. Halmos, *Naive Set Theory*, Springer-Verlag, NY, 1998.

P. R. Halmos and P. Halmos, *Problems for Mathematicians, Young and Old*, Mathematical Association of America, Wash., DC, 1991.

D. R. Hofstadter, *Gödel, Escher, Bach: An Eternal Golden Braid*, Harper Collins, NY, 1999.

C. R. MacCluer, *Industrial Mathematics*, Prentice Hall, Upper Saddle River, NJ, 2000.

C. R. MacCluer, *Boundary Value Problems and Fourier Expansions*, revised edition, Dover Publications, NY, 2004.

C. R. MacCluer, *Calculus of Variations*, Prentice Hall, Upper Saddle River, NJ, 2005.

F. R. Moulton, *An Introduction to Celestial Mechanics*, 2nd revised edition, Dover Publications, NY, 1970.

H. Weyl, *The Continuum: A Critical Examination of the Foundation of Analysis*, Dover Publications, NY, 1994.

Index

Note: References of the form "ex. m.n" refer to the exercise m.n. References of the form "(m.n)" refer to that equation.

Abel summation, 140
Abel's test, ex. 9.44
Absolute value, 20
Absolute area, 102
Accumulation point, 26
Acceleration, 68; centrally directed, (6.29)
Achilles, 134
Algorithm; bisection, 36; Newton, 63
Alternating series test, 141
Analytic function, 151
Angle, radians, 50
Anomaly; eccentric, ex. 4.14; mean, ex. 4.14
Antiderivatives, 47
Archimedian property, 14
Arclength, 109
Area; absolute, 102; (under a) curve, 102; (of a) sector, ex. 5.19
Average value, 104
Axiom of choice, 16

$B(p, q)$, *see* the beta function
Ball; center, 21; open, 21; radius, 21
Basin of attraction, exs. 6.4, 9.50
Beach-party problem, 68
Beamwidth (half-power), ex. 8.72
Bessel function, ex. 9.29
Beta function $B(p, q)$, ex. 8.37
Bijective function, 4
Bisection, the method, 26, 36
Black holes, 76
Boolean algebra, 1
Borel sets, 97
Boundary of a set, 22
Bounded from above, 13

C, the field of complex numbers, ex. 2.12
Cancellation law; additive, ex. 2.1; multiplicative, ex. 2.5
Cantor's diagonalization, ex. 1.25
Cardinality of a set, 5
Cartesian product, ex. 1.33
Cauchy; convergence, exs. 9.31–9.33; principal value, ex. 8.68
Cavalieri's; first law, ex. 8.33; second law, ex. 8.34
Center of a ball, 21
Central limit theorem, 116
Centroid, 107

Cesaro summability, ex. 9.30
Chain rule, (5.17)
Channels, ex. 6.59
Closed; Loop, ex. 6.7; set, 22, ex. 3.8
Compact set, 23
Comparison test; (for) integrals, ex. 8.49; (for) series, 138
Complementation, 2
Complement of a set, 2
Complete metric space, exs. 9.32, 9.48.
Completeness of **R**, 14, ex. 9.32
Component of a set, ex. 3.29
Composition of two functions, 4; associativity, ex. 1.13; commutative, ex. 1.20
Concavity, 76
Conditional convergence, 141
Connected set, 22
Constant function, 30
Constraint, 67
Continuity; pointwise, 42; uniform, 93
Continuous functions, 30; adding, 33; multiplying, 34
Continuous image; compact, 32; connected, 31
Continuous integrands, 92
Continuum hypothesis, 16; generalized, 16
Contraction mapping, ex. 9.48
Convergence; absolute, 140; conditional, 141; interval of, 146; (of a) sequence, 134; (of a) series, 136
Convolution, ex. 8.61
$\cosh x$, exs. 8.56, 8.57
Coulomb's law, exs. 8.66, 8.67
Countably infinite, 6
Cournot competition, ex. 9.51
Cover, open, 23
Critical point, 66

Δy, dy, 66
Darboux's lemma, ex. 5.29
Darboux sums, 89
Demand, 66
De Morgan laws, 2
Derivative, 44; applications, 60; concept, 44; (of) exponentials, (8.32); implicit, 62; (of) inverse functions, 48; (of) logarithm, (8.23b), ex. 8.20; partial, 79;

(of) power functions, (5.11); (of) rational power functions, 49; (of) trig. functions, 50; rules, (5.13)–(5.19)
Descartes, 62
Diagonalization, ex. 1.25
Difference quotient, 44, (6.1)
Differentials, (6.9), 79
Dirichlet's example, ex. 4.21, 90
Dirichlet series, ex. 9.54
Dirichlet's test, 140
Distance, 20
Domain of a function, 2
Duality of ∪ and ∩, 2

Electric power/energy, ex. 8.6
Ellipse, ex. 6.19
Elliptic integral, ex. 8.47
Energy; integral of power, ex. 8.6; kinetic, 77, exs. 6.37, 6.58; potential, exs. 6.37, 6.58
Epsilon-delta proof scheme, 36
Escape velocity, 76
Euler's formula, ex. 8.58
Exponent rules, (8.31)
Extrema, 46

Falling bodies, (6.15)
Fermat's principle, ex. 6.11
Field, axioms, 12; ordered, 12
First derivative test, 67
Fixed point, exs. 4.27, 9.48, 9.49
Finite set, 5
Fluid resistance, (6.17)
Folium of Descartes, 62
Force; external, 70; inertial, 70; (of) water resistance, 69
Forensics, ex. 6.56
Function, 2; analytic, 151; bijective, 4; composition, 4; constant, 30; continuous, 30; domain of, 2; graph of, 2; implicit, 62; injective, 3; nondecreasing, 48; nonincreasing, 48; polynomial, (4.15); range of, 3; root of, 32; rational, 43; ruler, ex. 4.22; surjective, 3; zero of, 32, 36
Fundamental theorem of calculus, 91

Γ(s), *see* Gamma function
Galileo's experiment, ex. 8.59
Gamma function Γ(s), ex. 8.36
Gaussian distribution, 116
Geometric series, (9.9), (9.33)
Graph of a function, 2
Ground state, ex. 8.45
Group, ex. 1.36; symmetric, exs. 1.37, 1.37

Hadamard's formula, (9.44)
Harmonic; alternating series, 141; motion, (6.16), (8.4); series, (9.13)
Heine-Borel lemma, 26
Hooke's law, 101
Hyperbolic functions, ex. 8.56

Implicit; differentiation, 62; function theorem, 62
Improper integrals, 113; Cauchy p.v., ex. 8.68; Laplace transform, ex. 8.62
Induction proof, 13, ex. 2.16
Inequality, 12; triangle, (3.2d), (3.5c)
Inertial force, 70
Infinite series, 136
Infinite set, 5; countably infinite, 6; uncountably infinite, 6
Infinitesimals, 65, 101
Inflection point, 77
Initial value problem, 69, ex. 6.54
Injective function, 3
Integral test, 139
Instantaneous rate of change, 67
Integration, 89; methods, 112; numerical, 118; (by) parts, 113; (by) substitution, 112
Integrability, 96
Integrable, 90, 92
Interior point, ex. 3.20
Intermediate value theorem, 31; informal version, 32
Intersection of two sets, 1
Interval, 21
Interval of convergence, 146
Inverse; additive, 12, ex. 2.2; function, ex. 1.9; image, 4, ex. 1.28; multiplicative, 12, ex. 2.6
IVT, intermediate value theorem, 31
Inverse function theorem, 48
Inverse square law, (6.34), (6.39)
Iteration, 37, ex. 4.27

Kepler's; equation, ex. 4.14; laws, I–III, 71
Kinetic energy, ex. 6.37, 108

Lanczos derivative, ex. 7.14
Laplace transform, ex. 8.62
Law of cosines, ex. 5.17
Least upper bound, see supremum
Lebesgue; integral, ex. 9. 56; measure, 97
Left hand rule, (8.51a)
Leibniz, ex. 8.69
L'Hôpital's rule, exs. 9.36–9.39
Limit; concept, 41; (of a) difference quotient, 44; (at) infinity, ex. 5.38; (of a)

product, (5.6b); (of a) quotient, (5.7);
(of a) rational function, 43; right hand,
ex. 5.33; (of a) sequence, 134; (of a)
sum, (5.6a)
Limit comparison test, 138
Limit superior, ex. 9.27
Lindelöf principle, 28
Linear (affine) function, 30
Linear approximation, 65
Logarithm, 110
Loop, closed, ex. 6.7
Lord Russell's paradox, ex. 1.24
Lower Darboux sum, (7.3a)

Main sequence star, ex. 6.8
mapping, *see* function
Mass, 70, 74, 106
Maximum altitude, (6.45)
Mean μ, (8.42a)
Mean value theorem, 47, 104
Metric, 20
Metric space, exs. 3.9–3.13; complete,
exs. 9.32, 9.48
Midpoint rule, 133
Moment, 106; first, 106; (of) inertia, 108;
second, 108
MVT, *see* mean value theorem

N, the natural numbers, 3, 13
Napier base e, (8.29)
n-th term test, 137
Newton's; (derivative) notation, (5.10);
law of cooling, ex. 6.55; method, 63;
rule, 70; trick, 75
Normal distribution, (8.43)

Observed fact, 74
ODE, *see* ordinary differential eqn.
Objective function, 67
One-to-one function, *see* injective, 3
Onto function, *see* surjective, 3
Open; ball, 21; covering, 23; interval, 21,
23; neighborhood, 21; relatively, 37
Optimization, 66
Ordinary differential equation, 69
Osculating circle, 60

Pappus; first law, ex. 8.11; second law,
ex. 8.12
Paradox, Russell's, ex. 1.24
Partial derivative, 79
Partial fractions, 113
Partition, 89
Period; orbital, 72; (of a) pendulum,
exs. 6.58, 8.73
Phase-locked loop, ex. 6.59

Pigeonhole principle, 5
Planar pendulum, ex. 6.58
Point; (of) accumulation, 26; boundary,
22; exterior, ex. 3.20; interior, ex. 3.20
Point chasing, a proof method, 2
Poisson flow, 116
Polar coordinates, (6.24)
Polynomial function, (4.15)
Population growth, 68, ex. 6.50
Positive numbers, 12
Potential energy, ex. 6.37, (8.3)
Power series, 145; interval of convergence,
146; radius R of convergence, 146, (9.44)
Power set, exs. 1.30, 1.32, 1.41
Preimage, ex. 1.28
Probability density function, 115;
Gaussian, see normal; normal, (8.43),
exs. 8.76–8.78; Poisson flow, 116;
Rayleigh, ex. 8.46

Q, the field of rational numbers, ex. 1.25
Quantum mechanics, 117
Quotient rule, (5.19)

R, the field of real numbers, 12
Radar, exs. 6.15–6.16
Rate of change, 67
Ratio test, 147
Rational numbers **Q**, ex. 1.25
Rational power functions, 49
Radius of a ball, 21
Range of a function, 3
Rayleigh density function, ex. 8.46
Reactor scram, 69
Relatively open set, 37
Revenue, 66
Riemann integral, 89
Riemann sum, ex. 7.11
Riemann zeta function, (9.17), (9.31)
Right hand rule, (8.51b)
Ring of integers, **Z**, 6, ex. 2.17
Root test, ex. 9.26
Rules of exponents, (8.31)
Russell's paradox, ex. 1.24

Satellites, ex. 4.14
Schröder-Berstein theorem, 5, ex. 1.40
Schrödinger; equation, (8.44), (8.45)
Schwartzchild radius, 76
Scientific method, 70
SCRAM, 69
Secant lines, 60
Second derivative test, ex. 6.12
Sensitivity, 65, 79
Sequence, 134; limit of, 134; main, ex. 6.8

Sequential compactness, 136
Sequential continuity, ex. 9.20
Sequential convergence, 134
Series, 136; alternating harmonic, (9.22); conditionally convergent, 141; convergence, 136; Dirichlet, ex. 9.54; divergence, 137; geometric, (9.9); grouping, 144; harmonic, (9.13); power, 145; reordered, 142
Set, 1; closed, 22; closure, ex. 3.23; empty set ∅, 1; finite, 5; infinite, 5; open, 21; power set, exs. 1.30, 1.32
sgn x, *see* Signum function
Shear strength, ex. 6.48
Signum function, ex. 5.33
Simpson's rule, (8.53), (8.56); error estimate, (8.56)
sinh x, exs. 8.56, 8.57
Snell's law, ex. 6.11
Snowplow problem, ex. 6.51
Speed control, ex. 6.7
Spring-mass system, 101
Squeeze lemma, ex. 5.22
Stablity, exs. 6.7, 6.59
Standard deviation σ, 129
Star, main sequence, ex. 6.8
Stationary state, 117, ex. 8.45
Statistics, 115
Stiffness, ex. 6.13
Submarine problem, 69
Subset, 1
Subspace topology, 37
sup, supremum, 13
Surjective function, 3

Tangent lines, 60
Taylor expansions, (9.58)–(9.61)
Taylor's theorem, ex. 5.20, (9.54)

Topology, 21; relative, 37
Topological space, ex. 3.28
Translate, ex. 4.1
Trapezoid rule, (8.52)
Triangle inequality, (3.2d), (3.5c)
Trigonometry, 50

Unbounded; integrand, 113; interval, 114
Uncountably infinite, 6
Uniformly continuous, 93
Union of two sets, 1
Universal gravitation, (6.38), 73
Upper Darboux sum, (7.3b)
Urey's problem, ex. 8.75

van der Waerden's example, ex. 9.53
Variable limits of integration, 95
Variance v, (8.42b)
Velocity, 67
Voltage; AC., ex. 8.6; DC, ex. 8.6; gain, ex. 8.72
Volume, 105; (of a) ball, exs. 8.4, 8.38; (of) rotation, 105, ex. 8.11

Water resistance, 69, (6.19)
Wave function, 117
Well ordering; (of) **N**, 13; Zermelo, 16
Weierstrass example, exs. 5.29, 9.53
Work, 100

Yagi antenna, ex. 8.72

$\zeta(s)$, Riemann zeta function, (9.17), (9.31)
Z, the ring of integers, 6, ex. 2.17
Zeno's paradoxes, 134
Zermelo's axiom of choice, 16

Milton Keynes UK
Ingram Content Group UK Ltd.
UKHW020703260624
444527UK00001BB/9